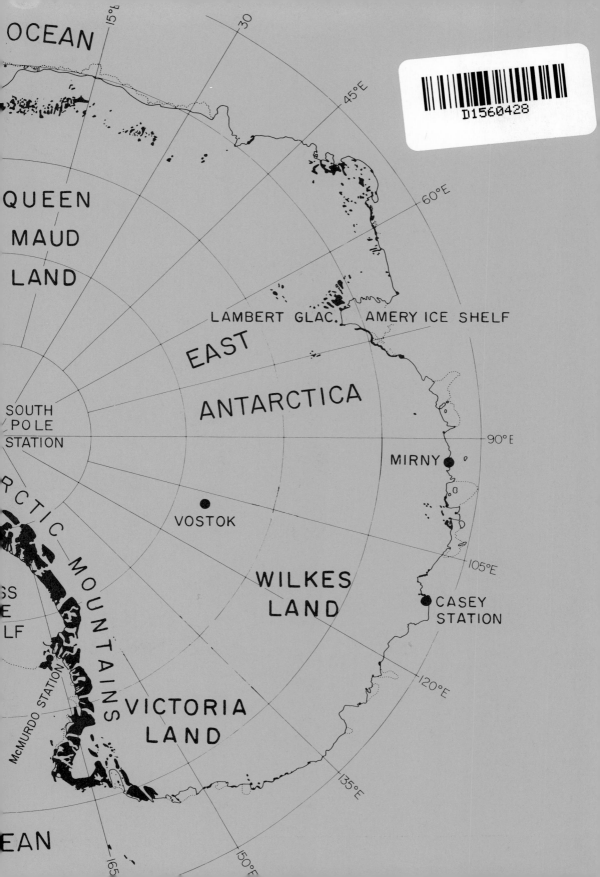

RADIOGLACIOLOGY

GLACIOLOGY AND QUATERNARY GEOLOGY

A Series of Books

RADIOGLACIOLOGY

by

V.V. BOGORODSKY

Arctic and Antarctic Scientific Research Institute, Leningrad, U.S.S.R.

C.R. BENTLEY

Geophysical and Polar Research Center, University of Wisconsin-Madison, U.S.A.

P.E. GUDMANDSEN

Technical University of Denmark, Copenhagen, Denmark

D. REIDEL PUBLISHING COMPANY

A MEMBER OF THE KLUWER ACADEMIC PUBLISHERS GROUP

DORDRECHT / BOSTON / LANCASTER / TOKYO

Library of Congress Cataloging-in-Publication Data

Bogorodskiĭ, V. V. (Vitaliĭ Vasil'evich)
  Radioglaciology.

  (Glaciology and quaternary geology)
  Translation of: Radioglîa͡tsiologiîa͡.
  Bibliography: p.
  Includes index.
  1. Radar in glaciology. I. Bentley, Charles R.
II. Gudmandsen, P. E. III. Title. IV. Series.
GB2401.72.R32B6413   1985   551.3'1'028   85–18283
ISBN 90–277–1893–8

Published by D. Reidel Publishing Company,
P.O. Box 17, 3300 AA Dordrecht, Holland.

Sold and distributed in the U.S.A. and Canada
by Kluwer Academic Publishers
190 Old Derby Street, Hingham, MA 02043, U.S.A.

In all other countries, sold and distributed
by Kluwer Academic Publishers Group,
P.O. Box 322, 3300 AH Dordrecht, Holland.

Originally published in 1983 by Gidrometeoizdat under the title
Radioglyatziologiya

Translated by V. Chebotareva

All Rights Reserved
© 1985 by D. Reidel Publishing Company, Dordrecht, Holland
No part of the material protected by this copyright notice may be reproduced or
utilized in any form or by any means, electronic or mechanical,
including photocopying, recording or by any information storage and
retrieval system, without written permission from the copyright owner

Printed in The Netherlands

DEDICATION

The English translation of this book is dedicated to the memory of Amory H. ('Bud') Waite, the pioneer in radar sounding of ice sheets, who made the first recorded long-distance transmission of radio waves through ice at Little America Station on the Ross Ice Shelf in the summer of 1955–56, when he recorded radio signals that had traveled one mile between snow pits. He also recorded the first bottom echo from the base of the Ross Ice Shelf in January 1957, although he did not recognize it as such until looking back later at his records. His first deliberate attempt at measuring ice thickness, at Little America Station in December 1957, was a discouraging failure. Only a month later, however, he successfully sounded ice up to 600 m thick south of Wilkes Station (now Casey Station). He also made the first airborne soundings in Greenland in the fall of 1959 and the first airborne survey in Antarctica in December 1961. In the summers of 1963 and 1964 he organized and coordinated the International Cooperative Field Experiment in Glacier Sounding, a unique venture in which the results of radar, seismic, gravimetric, and electrical measurements of ice thickness were compared directly along the same sounding lines.

Bud Waite initiated the second author's interest in radar sounding, and his generous help was instrumental in establishing a program of radioglaciology at the University of Wisconsin. That author is deeply grateful.

TABLE OF CONTENTS

PREFACE ... ix

CHAPTER 1: A BRIEF SURVEY OF THE EARTH'S ICE COVER AND METHODS FOR ITS INVESTIGATION ... 1

1.1. The Geographical Distribution of Glaciers ... 1
1.2. Impact of Ice on Geophysical Processes ... 6
1.3. The Seismic Method ... 7
1.4. Gravitational and Magnetic Methods ... 8
1.5. Radar Sounding ... 10
References: Chapter 1 ... 14

CHAPTER 2: STRUCTURE AND PHYSICAL STATE OF GLACIERS ... 16

2.1. Internal Structure of Glaciers ... 16
2.2. Stratification and Temperature Regime of Glaciers ... 20
2.3. Glacier Regimen ... 27
References: Chapter 2 ... 31

CHAPTER 3: ELECTROMAGNETIC WAVE PROPAGATION IN ICE ... 32

3.1. Crystalline Structure of Ice ... 32
3.2. The Electrical Properties of Monocrystalline Ice ... 35
3.3. Values of the Electrical Parameters ... 37
3.4. Pure Polycrystalline Ice ... 39
3.5. Impure Polycrystalline Ice ... 40
3.6. Snow ... 40
3.7. Propagation of Electromagnetic Waves in Glaciers ... 41
References: Chapter 3 ... 47

CHAPTER 4: EQUIPMENT FOR RADAR SOUNDING OF GLACIERS — 48

4.1. Analysis — 48
4.2. Estimation of Sounding Accuracy — 52
4.3. Equipment to Measure the Thickness of Cold Glaciers — 55
4.4. Equipment to Measure the Thickness of Temperate Glaciers — 58
4.5. Radars to Study Internal Structure and State — 59
4.6. Radars for Ice Movement Measurements — 62
4.7. Laser Technique — 62
4.8. Equipment of U.S.A., Denmark, and Great Britain for Radar Sounding of Cold Glaciers — 64
4.9. Equipment of U.S.A., Denmark, and Great Britain for Radar Sounding of Temperate Glaciers — 69
4.10. Specialized Equipment — 74
References: Chapter 4 — 76

CHAPTER 5: METHODS OF ACQUISITION AND PROCESSING OF DATA — 79

5.1. Radioglaciological Data Recording — 79
5.2. Digital and Photographic Recording of Radioglaciological Observations — 79
References: Chapter 5 — 83

CHAPTER 6: SCIENTIFIC RESULTS IN RADIOGLACIOLOGY — 84

6.1. Direct Measurements of Permittivity and Conductivity — 84
6.2. Electromagnetic Wave Speed and its Measurement — 93
6.3. Lateral Waves — 116
6.4. Absorption of Electromagnetic Waves — 120
6.5. Polarization Studies — 125
6.6. Ice Thickness and Subglacial Topography — 132
6.7. Subglacial Physiography and Geology — 159
6.8. Radar Sounding of Internal Layering — 164
6.9. Movement of Glaciers — 189
6.10. Nature of the Basal Interface — 216
6.11. Detection of Hidden Crevasses — 226
6.12. Estimation of Glacial Water Content — 233
References: Chapter 6 — 241

CONCLUSION — 250

SUBJECT INDEX — 253

PREFACE

Antarctica, the sixth continent, was discovered more than 160 years ago. Since then this large, mysterious continent of ice and penguins has attracted world interest. Scientific expeditions from various countries have begun to study the geographical and natural conditions of the icy continent. Systematic and comprehensive investigations in the Antarctic started in the middle of our century.

In 1956 the First Soviet Antarctic Expedition headed to the coast of Antarctica. Their program included studies of the atmosphere, hydrosphere and cryosphere. Thirty years have since passed. Scientists have unveiled many secrets of Antarctica: significant geophysical processes have been investigated, and a large body of new information on the Antarctic weather, Southern Ocean hydrology and Antarctic glaciers has been obtained. We can now claim that the horizons of polar geophysics, oceanology, and particularly glaciology, have expanded. Scientific investigators have obtained new information about all Antarctic regions and thus have created the opportunity to use the Antarctic in the interests of mankind.

Science in Antarctica was given a new impetus when the continent was declared a continent for peace and scientific cooperation. Since 1955 scientists from the U.S.S.R., U.S.A. and other countries have worked actively at many locations around the continent. Many international field experiments have had as their goals the study of the atmosphere, hydrosphere and ice. Participants of these international efforts never failed to exchange scientific results and information on the techniques and equipment used. Such exchanges have contributed greatly to the effectiveness of the research efforts.

Radar methods in glaciology have developed only recently. In 1955, V.V. Bogorodsky of the Arctic and Antarctic Research Institute and V.N. Rudakov of the Ulyanov (Lenin) Electrotechnical Institute, in Leningrad, developed the theory of electromagnetic wave propagation through glaciers, determined the range of radio transparency, and outlined the capabilities of pulse radars for glacier thickness measurements.

In 1956, V.V. Bogorodsky, V.N. Rudakov, V.V. Pasynkov and Z.I. Schwainstein received patents for their proposal to use remote-sensing radio methods for measuring the thickness of sea ice.

In 1957 the American polar explorer Amory Waite made the first deliberate radar sounding when he recorded an echo from the bottom of the Ross Ice Shelf.

In 1960, V.V. Bogorodsky and V.N. Rudakov (1960) published a paper entitled 'On the question of glacier thickness measurement by electromagnetic methods', in which the calculations of all radar parameters for radar sounding of 2 km-thick glaciers were given.

In 1964 the engineers V.A. Tulpin and I.I. Pronin carried out the first radar measurements of glacier thickness to be made by the Soviet Antarctic Expedition (SAE). These separate theoretical and field efforts of Soviet and US scientists gave birth to the radioglaciology of today.

The last 20 years have seen the rapid expansion of radioglaciological investigations carried out principally by the scientists of the Soviet Union, the United States and the United Kingdom in Antarctica; by Canadians in North America; by Danish scientists in Greenland; and by Soviet scientists of the Leningrad Arctic and Antarctic Research Institute (AARI) in the Soviet Arctic. Scientists from the U.S.S.R., U.S.A. and Denmark have been exchanging radioglaciological information for years. This exchange has proved extremely valuable: enabling the efficiency of radioglaciological studies in all ice provinces of the Earth to be increased. This international cooperation has contributed greatly to the development of techniques and equipment for remote sensing of ice thickness, for determining bedrock morphology, and for locating sub-glacial lakes. Radar techniques have enabled us to determine the rate of ice movement, to estimate quantitatively the solid discharge from the continent to the ocean, to measure effective ice temperatures, to reveal ice stratification down to great depths, to detect crevasses in ice, and to calculate the water content of glacial ice. One of the important achievements has been the establishment of a set of previously unknown electrical properties of the Antarctic ice sheet, which is regarded as a dielectric of unlimited size. Radar techniques are extensively used together with gravimetric and magnetic measurements.

The problems resolved and the new scientific achievements attained in the study of inland ice sheets, outlet glaciers, and ice shelves confirm the fact that the radar-sounding technique has been extensively used for investigations of the polar regions. All these important results suggest that the foundation of a new and promising technique in geophysics, i.e., radioglaciology, has been laid down. The authors decided that the time was ripe to summarize the results of long-term glaciological studies by means of modern, effective radiophysical methods. This job was a difficult one and many problems had to be overcome, but we were stimulated by the belief that our monograph covering radiophysical studies in glaciology would promote the development of new approaches to the study of glaciers in their relation to the world ocean and climate.

Airborne and satellite techniques for glacier studies should be further developed. This task can be successfully handled on the basis of international agreement for cooperation between scientists of different nations. International cooperation would expedite the acquisition of new data, which are required for both scientific and applied purposes.

The book is divided into six chapters. Chapter 1 contains general information on glaciers and the major ice provinces of the Earth, and on methods to investigate ice structure and state. Chapter 2 gives a review of the studies on structural and physical properties of Antarctic, Greenlandic and Arctic (Soviet Arctic) glaciers. Chapter 3 discusses electromagnetic wave propagation in ice, covering electromagnetic properties of ice monocrystals and of polycrystalline ice and snow. Wave scattering and reflection in the ice and at ice boundaries are also dealt with in this chapter.

Chapter 4 describes equipment used by scientists from the U.S.S.R., U.S.A., Denmark, and other countries to study glaciers. This chapter gives recommendations as to which radar to choose for the study of various glacier types. The chapter discusses ice-thickness measurements, bedrock morphology surveys and the capabilities of radar techniques to estimate water content in glaciers and to discover sub-glacial lakes.

Chapter 5 contains a discussion of recording and processing systems used at present for radiophysical observations. Chapter 6 comprises summaries of a large amount of new material that researchers have accumulated through the utilization of radiophysical methods of studying ice. The chapter includes information on the thickness of ice sheets and the morphology of their beds, the movement and temperature of glaciers and their internal layering and fracturing, the possibility of age-dating by radar sounding, the determination of the equivalent water masses of glaciers, subglacial lakes, and a few other topics.

In Chapter 1, sections 1.1., 1.2, and 1.5 were written by Bogorodsky; sections 1.3 and 1.4 by Bentley. In Chapter 2, sections 2.1 and 2.3 were written by Bentley; section 2.2 by Bogorodsky. In Chapter 3, sections 3.1–3.6 were written by Bentley; section 3.7 by Gudmandsen. In Chapter 4, sections 4.1 and 4.2 were written by Gudmandsen; sections 4.3–4.7 by Bogorodsky; sections 4.8–4.10 by Bentley. In Chapter 5, section 5.1 was written by Bogorodsky; section 5.2 by Bentley. In Chapter 6, sections 6.1, 6.3, 6.7 and 6.10 were written by Bentley; sections 6.4 and 6.12 by Bogorodsky; sections 6.2, 6.5, 6.6, 6.9, and 6.11 by Bentley and Bogorodsky; section 6.8 by Bentley, Bogorodsky and Gudmandsen. The Preface and Conclusion were written by Bogorodsky.

The authors address their book to a steadily growing company of polar scientists, men and women who are ready and willing to face the challenge of a stimulating career in polar investigation.

In thanking those who aided in the production of this book, the authors wish to acknowledge, in particular, V.A. Chebotareva, O.V. Boikova, and A.N. Sheremetyev, Leningrad, for special technical assistance, and M. Bentley, Madison, for editorial assistance.

The authors do not think that their book is flawless. They would greatly appreciate readers' comments and will take them into consideration in their future work.

CHAPTER 1

A BRIEF SURVEY OF THE EARTH'S ICE COVER AND METHODS FOR ITS INVESTIGATION

1.1. The Geographical Distribution of Glaciers

Glaciers are multi-year ice masses capable of movement. They form mainly from solid atmospheric precipitation in those areas of the Earth where there is not melting during a warm season and where rainfall is negligible.

Long-term accumulation of solid precipitation results in growth of the snow cover. Complex metamorphic processes, with time, transform the snow into firn, and the firn eventually into glacial ice. Snow is the major material for glacier formation, but ice can also form by the penetration and refreezing of melt water, forming congelation ice. The long-term accumulation of solid precipitation (snow, hail, hoar frost, freezing rain) is called glacier accumulation. The opposite process, glacier mass reduction, is called ablation. Both accumulation and ablation are affected by numerous complex physical processes. The rate of glacier accumulation and ablation varies with time and depends on the geographic position of a glacier.

Newly fallen dry snow with a density of $0.06\,\mathrm{mg\,m^{-3}}$ is in the form of ice crystals, hexagonal in shape. Wind moves the snow, breaking the snow flakes and compacting them. This process results in snow recrystallization. It sometimes happens that falling snow flakes are encrusted with a thin water film – such snow is called wet snow; its density normally ranges from 0.08 to $0.15\,\mathrm{mg\,m^{-3}}$.

There is also moist snow; its flakes, unlike those of wet snow have no regular crystal shapes. Moist snow usually falls in clumps, its density being more than $0.2\,\mathrm{mg\,m^{-3}}$. Any snow type undergoes a metamorphosis characterized by rapid changes of its physical properties. With time, snow consolidates into firn with its peculiar physical properties. Snow with density of $0.4-0.8\,\mathrm{mg\,m^{-3}}$ is considered to be new, transient ice (Kalesnik S.V., 1963) [4]. Firn with densities higher than $0.8\,\mathrm{mg\,m^{-3}}$ is called ice. The transformation of firn into ice is a process whose rate depends strongly on the temperature and the amount of solid precipitation. In the Zailiysky Alatau Mountains, for instance, it takes firn 1–2 years to become ice, whereas in the vicinity of Mirny Observatory, Antarctica, the process takes 190–200 years, and in inland Antarctica the process continues for about 1000 years (Kotlyakov V.M., 1961) [5].

The foregoing processes finally result in glacier mass accumulation and growth (in both extent and thickness). With the increase in glacier thickness pressure gradients form that cause plastic flow of the ice and glacier movement. If the ice mass does not move, it is called a multi-year ice cover (Kalesnik S.V., 1963) [4].

The mass, dimensions, and surface topography of a glacier are determined by geophysical conditions, glacier structure, glacier state, and bedrock topography.

Those glaciers whose surfaces essentially repeat the large-scale topography of their beds are called mountain glaciers. Vast glaciers whose surface relief does not depend on that of the bed, being determined mainly by their inner physical state and by the geophysics of their environment, are called ice sheets. Normally ice sheets are found in high latitudes covering vast areas. Their masses are enormous. The Antarctic and Greenland ice sheets are the largest on the Earth's surface. The Antarctic ice sheet covers the continent of Antarctica, whereas the Greenland ice sheet covers the large island of Greenland.

Continental and island ice sheets consist of ice domes, outlet glaciers and ice shelves. Frequently all these features form a single geophysical complex. The diameter of an ice dome often reaches thousands of kilometers, its thickness being about 3000–4000 m. Normally the highest domes coincide with rises in the bedrock, quite frequently the maximum thicknesses of such glaciers are found over deep depressions in the bedrock. Outlet glaciers are peculiar ice rivers flowing either over sloping beds (e.g., Lambert Glacier) or along narrow channels (e.g., Denman Glacier). Outlet glaciers provide for the ice discharge from inland areas to the continental periphery.

An ice shelf is a flat margin of the Antarctic ice sheet, mostly afloat. An ice shelf usually terminates in a steep barrier.

Mountain glaciers either fill depressions in the mountain relief or cover individual summits and plateaus, the former being called cirque glaciers. Their size is limited by that of a crater, corrie, or cirque. Glaciers covering valleys between mountains are called valley glaciers. They frequently occupy most of the river valleys in the mountains. Flat ridges and alpine plateaus are covered with dome-like glaciers. Flat vast uplands can be covered by the so-called plateau glaciers. Island glaciers (e.g., Franz-Joseph Land) differ from the ice sheets only in their size, having all morphological elements in common with the ice sheets.

The surface profile of large ice sheets resembles an ellipsoid, a shape that is typical of a viscous elastic body spreading under the force of gravity. This surface is known to have a number of features. The surface of the Antarctic ice sheet gradually rises towards its center. There are gentle hills and swells on the surface, their height diminishing inland. In the coastal zone, which is 60 km wide, surface features 25–30 km high and 6 km wide are quite common. The slope of some coastal zone irregularities reaches 0.25, the average being 0.02–0.05. The surface of the Greenland ice sheet resembles that of the Antarctic ice sheet.

Domes and outlet glaciers are the main morphological glacier types in the Soviet Arctic. Some of the glaciers of Franz-Joseph Land and Severnaya Zemlya are terminated by ice shelves with gently sloping surfaces. The largest domes are found on the Severnaya Zemlya glaciers. The domes in Severnaya Zemlya are characterized by a complete absence of crevasses, whereas the outlet glaciers on the same archipelago are heavily criss-crossed by crevasses and chaotic ice pile-ups. The outlet glaciers of the central islands in Franz-Joseph Land have the most irregular surfaces.

Dimensions, shapes, masses, inner structures, and states of glaciers change considerably from one climatic epoch to another. At present, the Earth's glaciers contain approximately 1.62 per cent of the fresh water of the planet (about $1.5 \times$

10^{18} tons), which is 0.004 per cent of its entire mass. It is apparent that the ice mass is very small in comparison with the Earth's mass. It would be, however, quite sufficient to cover the surface of the Earth by an ice layer about 60 m thick. If the ice melts (providing the water temperature remains 0°C), the level of the world ocean would rise 60–70 m and the ocean area would increase by 15 million km^2.

The existence and distribution of the ice cover is controlled by the geographic latitude of the accumulation limit and the bedrock elevation above sea level. Accumulation zones in polar regions need not be high with respect to sea level. In Antarctica, ice accumulation occurs even at sea level. In temperate and tropical zones, however, accumulation zones are found only at considerable heights in the mountains. There different locations of accumulation zones are mainly determined by thermodynamics, i.e., by the amount of atmospheric heat available for ice melting. Solar radiation and turbulent convection provide thermal energy to the ice. Direct and scattered solar radiation cause snow and ice melting on alpine glaciers. For arctic glaciers of lower height, radiant and turbulent-convective energy fluxes are roughly equal; scattered radiation becomes more important here due to a substantial amount of cloud, which is typical of the Arctic.

The physical conditions of antarctic glaciers differ from those in the Arctic. In summer, sunny weather prevails over Antarctica, so antarctic glaciers are subject to strong direct solar radiation. In Antarctica, glacier surfaces receive a large amount of radiative energy even on cloudy days. However, due to the high reflectivity of the ice surface, only a small part of this energy is used for melting. That is why the main mass loss of the Antarctic ice sheet is through solid discharge from inland areas to the continental periphery rather than through melting. Calving of ice shelves, flow of outlet glaciers and intense melting of ice at the periphery contribute greatly to this effect.

In summing up the foregoing information on the stages of glaciers and the physical phenomena occurring within them, it can be concluded that each glacier is characterized by three main processes that determine its structure and state: mass accumulation, movement due to gravity, and reduction of mass by ablation.

Ice covers large land areas north of 60°N and south of 50°S. In lower latitudes, mountain glaciers occupy small areas that are only fractions of a per cent of the total land surface. At present, ice sheets cover 16.2 million km^2, or approximately 11% of the land. This does not, however, include areas with seasonal ice cover, the sea-ice covered regions of the Arctic and Southern Oceans, iceberg-infested waters and permafrost regions. Considered together, all these areas of glaciation would cover 20% of the land. It is apparent that this extensive distribution of ice will affect all physical processes and all aspects of human activity. It is in this connection that large and developed countries like the U.S.S.R., U.S.A., Canada, Sweden, Norway, Denmark, and some others are strongly affected by ice.

Tables I and II illustrate the geographic distribution of glaciers and their locations on different continents. As is seen in Table II, the Antarctic and Greenland ice sheets are unique in their size – they almost completely cover the land masses of Antarctica and Greenland, and contain the major portion of the Earth's ice mass.

Tables I and II do not present the information on the mass and area of the

TABLE I
Latitudinal ice distribution

Latitude	Glaciers, per cent of the land area	Sub-surface ice, per cent of the land area	Sea ice, per cent of the ocean area	All ice forms in per cent of the total area of the land and ocean
90–80°N	68.5	31.5	92.5–97.3	93.2–97.6
80–70	35.8	64.2	66.1–86.9	79.1–90.8
70–60	4.5	80.3	22.8–65.2	67.7–79.0
60–50	0.30	45.2	7.6–35.1	29.3–41.1
50–40	0.07	5.7	4.1–13.1	5.0–9.1
40–25	0.47	1.7	0	0.91
25°N–30°S	0.00	0.00	0	0
30–35	0.08	0.01	0	0.01
35–50	0.84	0.10	0	0.04
50–60	10.9	0.8	0.0–11.1	0.19–17.8
60–90	99.93	0.07	24.5–84.2	55.2–90.6

TABLE II
Areas of the glaciers on different continents

Region	area of the glaciers		
	km^2	per cent of the continental area	per cent of the total glacier area
Antarctica	13 914 000	99.93	85.81
Greenland	1 802 400	82.45	11.11
North America	217 300	0.98	1.34
Asia	137 100	0.33	0.85
Europe	117 000	1.01	0.72
South America	26 400	0.14	0.16
Australia	1000	0.03	0.01
Africa	12	0.00004	0.0001

glaciers in the Soviet Arctic. Their area is 55 865 km^2. They are located on Franz-Joseph Land, Novaya Zemlya, Victoria Island, Ushakov Island, De-Long Island, and Severnaya Zemlya, covering almost 54% of the area of those islands. Table III shows areas of individual glaciers, degree of glaciation, glacier thickness, and mass.

In various geological epochs the mass of ice and the area glaciated has varied substantially. During the last glaciation 20 000 years ago, for instance, ice covered much more land than at present. In the Quaternary, glaciations waxed and waned. The last maximum (glacial epoch) occurred about 10 000 years ago. During that time the ice volume was estimated to be 89 million km^3, covering an area of 38 million km^2. Thus the glaciers of that period were three times larger in mass

TABLE III
Characteristics of the Soviet arctic glaciers

Region	Glacier area km²	Area of ice-free land km²	Degree of glaciation %	Mean ice thickness (estimated) m	Approximate ice volume km³	Ice mass (approx) tonnes
Franz-Joseph Land	13 690	2400	85.1	100	1370	12×10^{11}
Novaya Zemlya	24 300	57 879	29.6	280–300	6800	612×10^{10}
Severnaya Zemlya	17 470	19 300	47.6	200	3500	3×10^{12}
Victoria Island	5.2	0.1	98.0	50–60	0.33	3×10^{8}
Ushakov Island	323	–	100.0	100	10	9×10^{9}
De-Long Islands	77	140	35.5	90–100	7	6×10^{9}
Total	55 865				11 700	

and 5 times larger in the area they covered than those of the present (Monin A.S., 1977) [6]!

Glaciations also change on a shorter time-scale than a geological epoch — sometimes a large increase or decrease takes only one or a few years. Short-term changes of ice mass are caused by the variations in accumulation and ablation. Precipitation is known to be subject to short-term aperiodic changes. Variations in ice mass that result from changes in the rates of mass accumulation, ablation, and evaporation lag very little behind the changes in the geophysical environment. Longer time lags, however, are involved in increases of glacier thickness and glacial extent associated with the same environmental changes.

Changes of external geophysical conditions, in turn, change the mass of glaciers, their velocities and their areal extent, each glacier responding differently, due to varying relaxation times, to the changes of physical conditions. Changes in ice mass are slower in inland areas, and faster in regions with a humid marine climate. Govorukha (1968) [3] discussed evidence on the reduction of glacial ice volume during the period from the late 19th century to the middle of the present one. This process of glacier retreat is a common feature of all geographic areas. Govorukha (1968) [3] cites the following facts confirming his conclusion: Swiss glacier mass diminished by 10% from 1922 to 1944 with a 25% decrease in area; the glaciers of Franz-Joseph Land diminished by 8% from 1930 to 1944. In the 20th century a slow retreat of glacial ice has been observed in north-eastern Greenland, in the Suntar-Khayata Range, in Central Asia, and in Antarctica.

This ice retreat seems to continue in most regions of the Earth. It should be noted here that there is some evidence of a mass increase in some glaciers since the 1950s (e.g., glaciers in the mountains of the northwestern United States, southwestern Canada, and Alaska). The same is true of glaciers in Central Asia, the Mediterranian Alps, and the Pamirs. Thus, a conclusion can be made on the completion of the period of glacier retreat in this century. One should bear in mind, however, that the foregoing speculations on glacier retreat and advance are of hypothetical character, since no experimental confirmation is available. It is, however, obvious that glaciers are sensitive indicators of climatic change.

Quantitative glacier mass changes and glacier structure and state can be studied successfully by radioglaciological methods.

1.2. Impact of Ice on Geophysical Processes

Temporal variations of ice mass have already been mentioned. Ice ages are known in the Earth's history. During these periods ice sheets formed and their mass significantly increased. Drastic climate changes occurred in these periods resulting in changes in the Earth's biological environment (flora and fauna). During interglacials the climate is believed to have been warm. Tillites (ancient moraine deposits laid down by the ice) are good evidence that past glaciations covered vast areas. The tillite morphology shows that one of the centres of ice spreading was located in East Antarctica. It is from here that the ice of Gondwanaland flowed across the southern parts of Australia, Africa, and South America. Large ice streams spread northward from Africa and India.

The glaciation of the Northern Hemisphere was 13 times larger than the present one. The ice variations are believed to cause sea level changes in the world ocean of approximately 85–120 m, and formation of terraces on sea coasts. In the ice ages the World Ocean water level decreased significantly, exposing much of the continental shelves, and forming land bridges connecting Europe to the British Isles, and Chukotsky Peninsula to Alaska. (Monin, 1977) [6].

The ice mass growth resulted in land subsidence into the Earth's mantle. When the ice melted, the land, freed from its weight, rose. This rise for the past 9000 years has been 250 m near the former glaciation center in the northern Bothnia Gulf; the process continues at a rate of 1 cm per year and in the future it will reach another 200 m (!). Alternate land and water level rises would alternately separate the Baltic Sea from the North Atlantic and join them together again; these changes in physical conditions would in turn greatly alter the salinity structure of the Baltic Sea.

The recent Little Ice Age, known to have continued approximately from 1430 to 1850, was a result of climatic cooling. The cold maximum occurred in the 15th to 17th centuries. The surface water temperatures in the North Atlantic and Sargasso Sea were higher than the present ones, while the surface water temperatures around Iceland were lower than they are now (Monin, 1977) [6]. This reduced the heat flux from the ocean to the atmosphere in the trade wind area. These conditions in turn weakened the atmospheric circulation in the middle latitudes and caused a sharp surface water temperature rise in the Sargasso Sea and a surface water temperature drop around Iceland. Warming in the first half of the 20th century strongly affected the oceans, the Arctic Ocean in particular. The extent of sea ice in the Barents Sea then decreased by 20%, and the limit of iceberg occurrence around Antarctica shifted 1700 km closer to the sixth continent.

Apparently, variations of ice mass bring about significant changes in the physical state of all components of the Earth's surficial covering. Other changes are also known: changes of local landscapes, weather, the amount of water in lakes and rivers, of sea water structure (due to ice movement), ice cap extent and solid discharge from the ice sheets into the ocean. Thus, ice of all types and its variability directly and significantly affect human life and activity. Glaciers are enormous depositories of the purest fresh water; they are also huge refrigerators of the Earth.

A conclusion inevitably rises in the mind: man should study comprehensively all types of ice, and in particular watch the ice sheets, those unique natural relics, studying them with modern techniques and technology that do not interfere with their evolution.

Remote sensing is believed to meet perfectly this important requirement of our days. In the succeeding chapters of this book the main physical methods used currently for the investigation of the Earth's glaciers are discussed.

1.3. The Seismic Method

Prior to the development of the radar sounding technique, the principal means of determining the thickness of ice bodies was seismic reflection shooting. The shooting is done in a hole found or drilled into the ice. The waves generated by the shot are reflected and refracted by sharp changes in the acoustic impedance, and return to the surface where they are detected by geophones spread out on the glacier surface. The travel times of different wave types from the shot to the geophones is accurately recorded. By analyzing the amplitude, phase, and frequency of these waves it is possible to determine the depths and slopes of the reflecting interfaces and the acoustic wave velocities in the overlying and underlying media.

The principal limitation to the accuracy of the seismic reflection method is the uncertainty in the speed of propagation of seismic waves through the ice – an uncertainty which never exceeds ±5% and is often much less. Furthermore, both the uncertainty in the wave speed and the difficulty of detection of seismic reflections generally diminish as the thickness of the ice increases, so that the reflection method is particularly well suited for use on the major ice sheets. On the other hand, as the ice becomes thinner and the reflection technique becomes increasingly unsatisfactory, the refraction method becomes easier to use; the two seismic techniques are thus complementary in their application.

Powerful and useful as the seismic techniques are, they have a number of disadvantages. Perhaps the most important is that, more than for any other method, a skilled operator is necessary for successful results. Another disadvantage is the need to use motor-driven drilling equipment, since the depths required are beyond the range of a hand auger. The more complex methods may easily require two hours for a single reflection shot, including the time needed to lay out the geophone 'spread', drill the shot hole, and pick up the spread again. Even with the simplest system, nearly an hour is needed to complete one measurement. The total weight that must be transported from place to place is at least 100 kg, and usually much more. The final major disadvantage of seismic methods is that they must be carried out from a stationary position on the glacier surface. This means that transport across the glacier or ice sheet must be provided, or an aircraft capable of landing on snow of ice must be employed, if a survey of ice thickness is to be carried out. Operation in crevassed regions is thus severaly limited, if not impossible.

Despite the cumbersomeness of seismic methods, their greatly superior accuracy compared to all others prior to the development of radar sounding, combined with their effectiveness in revealing both the internal structure of the glacier itself and the geological structure of the glacier bed, has led to their widespread application. Their use on alpine glaciers goes back to 1926 (Mothes, 1926 [14]) and, in Greenland, to Brockamp's single measurement with the Wegener Expedition in 1929 (Brockamp et al., 1933 [10]). The reconnaissance surveying of major ice sheets began in Greenland in the 1940s (Joset and Holtzsherer, 1953 [12]) and in Antarctica about 1950 (Imbert, 1953 [11]; Robin, 1953 [15]). In the succeeding years, extensive work has been carried out on the ice sheets and on alpine glaciers throughout the world.

Even with the advent of radar sounding, the value of seismic sounding on ice remains, for several reasons. First, seismic methods provide information from beneath the bottom boundary of glaciers, whereas radar soundings cannot, except in very unusual circumstances. This makes seismic soundings particularly important on ice shelves, where they yield the depth of the electromagnetically impenetrable sea water beneath the ice. Second, and more generally important, they yield information that complements that obtained from radar measurements. Since seismic wave velocities increase with depth in the firn zone, whereas the reverse is true for electromagnetic wave velocities in the same region, seismic methods lead to a determination of the variation of density with depth, which is impossible using radar techniques and yet is needed for the accurate calculation of ice thickness. Conversely, the accurate determination of ice thickness by the radar technique permits the interpretation of seismic reflection times in terms of wave velocities, from which, in turn, the crystalline structure of the deep glacier ice can be deduced. This effective use of seismic and radar techniques together is important to glaciology, and is discussed further in Chapter 6.

1.4. Gravitational and Magnetic Methods

Measurements of the variation in strength of the Earth's gravitational and magnetic fields can be used to measure ice thickness.

The gravitational method is by far the more important of the two methods. The gravitational field strength at any point depends on the integration over all space of the mass in each element of space divided by the square of the distance between that element and the point of observation. Thus, the gravitational effect of any irregularity in the Earth depends on its distance from the observer and on the density contrast associated with the irregularity. The large density contrast between rock and ice (generally a ratio of about three to one) makes the gravity field relatively sensitive to variations in subglacial topography, and its measurement therefore useful for the calculation of ice thickness, assuming that some means are provided for taking into account the effect of more distant mass variations.

The gravitational method has one major advantage: ease of use. The total weight of the gravimeter package is only 10 to 20 kg, which provides extreme portability – measurements can easily be made by a single observer on foot or on skis. Its main disadvantage is that it is substantially less accurate than measuring the travel time of seismic or electromagnetic waves.

The first limitation on accuracy stems from the rapid change in gravity with elevation above sea level, i.e., with distance from the mass of the Earth itself – a 10 m change in elevation is roughly equivalent to a 50 m change in ice thickness. Although normally only the difference in elevation from a given base station, at which the ice thickness is known, is required, the crucial importance of accurate elevation measurements is apparent.

Secondly, the 'background' field from mass variations that are not related to the ice thickness changes must be determined. There are basically two different ways of correcting gravity observations for non-glacial effects: one, to measure the background field directly, and two, to provide ice thickness values by some more direct means (seismic, radar sounding, drilling) at a number of calibration points.

The first method is applicable to valley glaciers, small ice caps, and, to some extent, the marginal areas of larger ice caps and ice sheets, where the gravity measurements can be extended beyond the edges of the ice cover. A map of the gravity field can be constructed from the observations off the ice; interpolation yields the background field in the glaciated region. Differences between the interpolated background field and the values of gravity actually measured are then attributed to variations in ice thickness.

The accuracy of this method depends strongly on the local situation. Where the regions around the ice cover are easily accessible, the subglacial topography is subdued, and the geological structure beneath the ice cover is a simple continuation of that in the surroundings, accuracies of a few tens of meters in ice thickness are attainable. However, changes in gravity due to the geology could introduce serious errors if they are misinterpreted as part of the glacial effect.

Whenever possible, therefore, independent measurements of ice thickness should be obtained, at least at a few points. Even a single sounding in the middle of a glacier greatly improves the accuracy of the gravity analysis. On the major ice sheets and on most ice caps, where distances from the margins are too great to permit any valid estimation of the background gravity, such soundings are essential for success.

The gravitational effect of a subglacial topographic block depends primarily on its total mass, not its shape. Thus large errors in the evaluation of the local ice thickness are possible when the subglacial topography is rough and the ice is thick.

The first gravitational glacier soundings were made in Greenland in 1948 (Martin, 1948) [13]. Since 1950, the gravitational method has been used for Antarctic glacial investigations, mainly by combining closely-spaced gravimeter observations with more widely-spaced seismic soundings. The analysis of a large number of measurements at seismic sounding sites in Antarctica has led to an estimated average standard error from all effects of about ± 300 m, i.e., $\pm 15\%$ of the average ice thickness (Bentley, 1964, [9]). This error value can be taken as being representative for other regions, although, of course, a substantially greater or lesser accuracy may characterize a particular situation.

The gravitational method is useless for measuring the thickness of floating ice, but it can be useful in determining the depth to the underlying sea floor. The advantages and disadvantages are essentially the same as outlined above, except that the uncertainty in the ice surface elevation is greatly reduced.

The magnetic method has a much more limited applicability for ice thickness determinations, and consequently is little used. Analysis of magnetic anomalies leads to an estimate of the maximum depth to magnetically active rocks, i.e., rocks that either have a high susceptibility and thereby modify the main magnetic fields of the Earth, or have their own inherent magnetization that produces an additional field superimposed on that of the Earth. In either case, the relevant rocks are almost always igneous, and usually mafic. If there are layers of sedimentary or metasedimentary rocks beneath a glacier, the magnetic method can reveal only the total thickness of the ice and those layers together. Thus glacier thickness itself can be estimated only when there is good reason to believe, from independent geological evidence, that magnetically active rocks closely underlie the ice.

The great advantage of the magnetic method is that measurements can be made from an aircraft. Thus, in the proper setting, and especially where it is useful to know the maximum possible thickness of the ice, the rapidity with which an aeromagnetic survey can be conducted may make it worthwhile. Most likely, flights over a glacier would be made only as part of a more general survey that had some other primary purpose.

1.5. Radar Sounding

Radar sounding (echo sounding by pulsed radio waves), similarly to seismic sounding, is an active remote-sensing method. However, radar sounding of glacier characteristics can be measured from aircraft and from moving surface vehicles. Thus, in practice, it is very different from the seismic method, since it enables information about the ice sheet to be obtained continuously.

The physical basis of radar sounding is as follows: a short electromagnetic pulse is emitted by an antenna mounted on a platform moving over the glacier surface. The pulse penetrates the glacier and is reflected by inhomogeneities in the ice and by the bedrock, and an echo returns to the antenna. The ice thickness, depths of reflecting interfaces, and mean glacier temperatures and velocities can be determined by the analysis of the pulse delay time in the ice. The phase and amplitude characteristics of the echo should also be known.

The radar method has become one of the major tools used extensively by glaciologists all over the world. This method cannot be used successfully without an understanding of how the electromagnetic signal propagates through the ice and what the physical properties of the signal are.

Glaciers, as geophysical objects, have a very complex structure controlled by the physical conditions of their formation, such as precipitation cycles, climate and weather changes, and snow and firn metamorphism. A glacier moving over the bedrock interacts with it, entrapping rocks that penetrate the ice, forming moraine layers. Irregular glacier movement disturbs the continuity of the ice, resulting in the formation of crevasses, fractures, and voids. Thus, glacial ice has an inhomogeneous structure that is due not only to variations in the ice itself, but also to moraine intrusions, air bubbles, and sometimes even liquid water. Ambient air temperatures, solar radiation, and the thermodynamic interaction with the underlying bedrock determine the temperature distribution in the glacier.

Glacial ice is a dielectric; as such; its electrical properties can be specified by a (non-dimensional) complex relative permittivity

$$\varepsilon = \varepsilon' - i\varepsilon'', \tag{1.1}$$

and a complex relative magnetic permeability. Since it is non-magnetic, its permeability is simply 1.

In general, the dielectric permittivity depends on the frequency of electromagnetic oscillation. Since ice absorbs electromagnetic energy, the following equation is more convenient to use than (1.1):

$$\varepsilon = \varepsilon'(1 - i\tan\delta),$$
where $\tag{1.2}$
$$\tan\delta = \varepsilon''/\varepsilon' = \sigma/(\omega\varepsilon_0\varepsilon').$$

Here, σ denotes the conductivity, ω the circular frequency (rad s^{-1}), and $\varepsilon_0 = (1/36\pi) \times 10^{-9}$ F m^{-1}, the dielectric permittivity in vacuo.

For glacial ice, $\tan\delta \ll 1$. From the solution of Maxwell's equations it is known that a harmonic plane wave propagating along the z-axis in an absorbing homogeneous medium is described by

$$E(z) = E_0(z)\exp\left(i\omega\left(t - \frac{z}{c}\right)\right), \tag{1.3}$$

where $E_0(z)$ is the complex wave amplitude, t is the time and c is the speed of electromagnetic wave propagation in the medium. For ice,

$$c_i = \frac{c_0}{\sqrt{\varepsilon'}}, \tag{1.4}$$

where $c_0 = 2.988 \times 10^8$ m s^{-1} is the speed of wave propagation in vacuo.

Thus, the time of electromagnetic wave propagation through an ice layer (of thickness h) and back is

$$\Delta t = \frac{2h}{c_i} \tag{1.5}$$

Knowing the speed of the electromagnetic wave and measuring the time it travels in the ice (1.5), it is possible to calculate the ice thickness.

The electromagnetic wave loses energy as it travels through the ice, so the returning echo is smaller than the original pulse. The total change in power of the signal during its travel to the bedrock and back to the receiver, N_Σ, can be described by

$$N_\Sigma = N_G + N_R + N_\phi + N_D + N_A + N_P, \tag{1.6}$$

where N_G denotes geometrical spreading losses, N_R losses due to reflections from interfaces, N_ϕ changes in signal strength due to focusing effects (may be positive or negative) – called the focusing factor or refraction gain, N_D losses due to scattering, N_A losses due to signal absorption in the ice, N_P apparent losses due to rotation in polarization of the received signal relative to the orientation of the receiving antenna. The quantitative estimation of the total signal attenuation results in

$$N_\Sigma = -20\log(G\lambda) + 20\log h + N_A(T)h + 40.7, \qquad (1.7)$$

where G denotes the antenna gain coefficient, λ the wavelength, h the glacier thickness, and $N_A(T)$ the specific absorption, which depends on the glacier temperature T (see 6.4). $N_A(T)$ can be calculated using

$$N_A = 8.68 \frac{\omega}{2c_0} \sqrt{\varepsilon'} \, h \tan\delta \qquad (\text{dB m}^{-1}). \qquad (1.8)$$

The quantity 40.7 dB in (1.7) consists of signal changes due to scattering, focusing and depolarization, and can vary depending on the conditions.

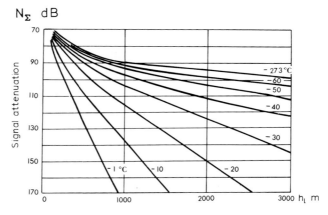

Fig. 1. Estimated dependence of total signal attenuation on the ice thickness for different absorption temperatures.

Figure 1 shows curves of total signal attenuation versus ice thickness at various temperatures. The calculations were made using (1.7). From the curves of Figure 1, the value of N_Σ can be determined. Knowing the receiver's sensitivity, it is then possible to select the transmitted power needed for ice thickness measurements.

Reflection and refraction of electromagnetic waves at the interfaces between two or more media are dealt with in detail in Chapter 3. Here, the equations for the reflection and transmission coefficients are presented for a case of vertical sounding by a plane wave incident on a planar interface between two homogeneous, isotropic media (Figure 2, curves 1 and 2). The reflection coefficient is

$$R_{1,2} = \frac{E_1}{E_0} = \frac{\sqrt{\varepsilon_1} - \sqrt{\varepsilon_2}}{\sqrt{\varepsilon_1} + \sqrt{\varepsilon_2}}, \qquad (1.9)$$

and the transmission coefficient is

$$T_{1,2} = \frac{E_2}{E_0} = \frac{2\sqrt{\varepsilon_1}}{\sqrt{\varepsilon_1} + \sqrt{\varepsilon_2}}, \qquad (1.10)$$

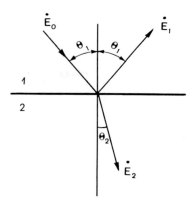

Fig. 2. Diagram of electromagnetic wave propagation through ice.

where E_0, E_1, E_2 are the complex amplitudes of incident, reflected, and refracted waves, respectively. Note that $T_{1,2}$ is always given by

$$T_{1,2} = 1 + R_{1,2} \tag{1.11}$$

Phenomena associated with wave propagation through the ice (time delay, wave dissipation, reflection from the bedrock and inner layers) constitute the physics involved. Equations (1.1)–(1.11) make possible a quantitative estimation of the physical phenomena accompanying electromagnetic wave propagation in glacier sounding.

The electrophysical properties of glacial ice vary in space considerably (Bogorodsky et al., 1977) [1]. The variation of electrical parameters between different layers and the boundary geometry of those layers are the most important features for radar sounding of glaciers. Note that the largest changes of E occur within the body of the glacier and are caused by changes in ice density, temperature and content of liquid water.

As the simplest case, a glacier may be described by a single homogeneous layer (Figure 3(a)). In such a model, the glacier is assumed to be a flat layer of unlimited horizontal extent, and to exhibit no significant changes in ε along the z-axis.

More generally, an ice sheet (or glacier) should be described using a multilayered model (Figure 3(b)); the medium underlying medium 1 consists of homogeneous layers having flat interfaces and thicknesses h_2, h_3, ... h_n, respectively, and ε changes from layer to layer either in discrete steps or exponentially. This model can be used for considering signal scattering or reflection from glaciological structures, such as snow, firn, and ice layers, ice layers with higher densities, layers containing moraine, etc. Analytical expressions for the reflection coefficient and other necessary parameters, discussed by Nikolsky (1973) [7], make it possible to determine the input impedance and reflection coefficient of a layer. A rough estimate of the radiowave scattering in a heterogeneous layer also can be obtained using a model described by Nikolsky (1973) [7].

Radar sounding of glaciers is carried out using short radio pulses – echoes from

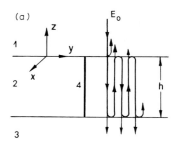

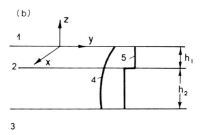

Fig. 3. Single-layer (a) and multi-layer (b) ice model: 1, 2, 3 – air, ice, and bedrock; 4, 5 – curves of permittivity variations; h_1 – snow–firn layer thickness; h, and h_2 – ice thickness.

glacier boundaries are received and the time delays between the maxima of their envelopes are measured.

The pulse duration to be used is determined by the system resolution required, i.e., it depends on the distance between the interfaces to be measured. Thus, if the thickness resolution is 10 m, the duration of the sounding pulse should not exceed 0.1 μs. When the sounding of ice sheets is carried out it is easy to choose a pulse length, τ which meets the requirement $\tau \ll \Delta t$. The choice of operating frequency for a particular purpose is determined by the conditions of wave propagation in the ice, by the desired rise time of the sounding pulse, and by the limitations of the platform upon which the system is mounted. Optimization of the carrier frequency is required to obtain the required power. A decrease in operating frequency and an increase in power results in an increase of the sounding depth. It should be noted, however, that a decrease in the operating frequency makes the resolution worse and the antenna larger and more difficult to mount on the carrier.

References: Chapter 1

[1] Bogorodsky, V.V., Kozlov, A.I., Tuchkov, L.T. *Radio emission of the Earth cover*. Leningrad, Hydrometeoizdat, 1977, 224 pp.
[2] Brekhovskykh, L.M. *Waves in stratified media*. Moskow, Science, 1973, 344 pp.
[3] Govorukha, L.S. 'Aspects of the glaciological study of the Soviet Arctic and the main peculiarities of its glacial regime'. *Trans. Soviet Geographical Society*, 1968, **100** (1).
[4] Kalesnick, S.V. *Essays in glaciology*. Moskow, Geographiz, 1963.

[5] Kotlyakov, V.M. 'Snow cover of Antarctica and its role in the present glaciation of the continent'. *Glaciology*, 1961, No. 7, 246 pp.
[6] Monin, A.S. *History of the Earth*. Leningrad, Science, 1977, 228 pp.
[7] Nickolsky, V.V. *Electrodynamics and radio wave propagation*. Moskow, Science, 1973, 608 pp.
[8] Finkelshtein, M.I., Mendelson, V.L., Kutev, V.A. *Radar sounding of stratified Earth covers*. Moskow, Soviet Radio, 1977, 174 pp.
[9] Bentley, C.R. 'The structure of Antarctica and its ice cover'. In: *Research in Geophysics*. Vol. 2, Massachussetts Institute of Technology, 1964, pp. 335–389.
[10] Brockamp, B., Sorge, E., Wölcken, K.K. *Wissenschafliche Ergebnisse der Deutschen Grönland-Expedition Alfred Wegener 1929 und 1930/31*. Leipzig, 1933, Bd. 2, pp. 1–160.
[11] Imbert, B. 'Sondages seismiques en Terre Adelie'. *Ann. Geophys.*, 1953, **9**, 85–92.
[12] Joset, A., Holtzscherer, J.J. 'Etude des vitesses de propagation des ondes seismiques sur l'Inlandsis du Groenland'. *Ann. Geophys.*, 1953, **9**, 330–344.
[13] Martin J. 'Rapport preliminaire de la campagne preparatoire au Groenland. Serie scientifique'. *Publ. des Expeditions Polares Francaises*, No. 5, Gravimetrie, 1948, 28–41.
[14] Mothes, M. 'Dickenmessungen von Gletschersis mit seismischen Methoden'. *Geolog. Rundschau*, 1926, **6**, 397–400.
[15] Robin, G. de Q. 'Measurements of ice thickness in Dronning Maud Land, Antarctica'. *Nature*, 1953, **171**, 55–58.

CHAPTER 2

STRUCTURE AND PHYSICAL STATE OF GLACIERS

2.1. Internal Structure of Glaciers

DENSITY STRUCTURE

Within the first year, the snow flakes falling on a glacier surface are altered to sub-rounded grains, forming firn. Thereafter, in a dry snow zone, where the surface temperature remains below the melting point throughout the year, there are three steps in the metamorphic process of densification. In the first and most rapid step, the grains rearrange themselves with little change in size or shape until they are as tightly packed as they can get. At this point the maximum packing density of about $0.55\,\mathrm{mg\,m^{-3}}$ has been attained. For further densification, the process slows to modification of the grains by means of sintering – mass transfer between the grains, principally in the vapor phase, recrystallization by molecular diffusion, and plastic deformation of the grains. Grain modification continues until the permeability of the firn has been reduced to zero, which occurs at a density of about $0.83\,\mathrm{mg\,m^{-3}}$. At this point the firn has been transformed, by definition, into glacier ice. Further densification then proceeds by a still slower process – bulk compression; here squeezing leads to a reduction in the pore size. The final result is ice that contains air bubbles, but that nevertheless is able to attain a density very close to that of single-crystal ice. At pressures of the order of $10\,\mathrm{MN\,m^{-2}}$, the gas is driven into the lattice of the ice and the bubbles disappear.

The whole process of densification occurs more rapidly if there is any melting in the upper layers at any time. Water increases the rate at which grains become rounded by melting their extremities, increases the average grain size (and thus the density) by preferentially melting smaller grains, accelerates packing by lubricating the grains, and fills in the air spaces by refreezing. Ice lenses commonly occur, and in an extreme case the entire snow layer may be transformed to ice in a single summer.

The depth to solid ice, even in the dry snow zone, will vary greatly from place to place. The colder the temperature and the greater the snowfall rate, the greater the depth to solid ice. In central East Antarctica the depth to ice of density $0.9\,\mathrm{mg\,m^{-3}}$ is approximately 170 m. In central Greenland and West Antarctica it is about 120 m, on the Ross Ice Shelf it is about 70 m, and in other places it may be substantially less. Where melting does occur, of course, the depth of melting can be very small – as little as the thickness of the seasonal snow layer. Plots of density vs. depth in the dry snow zone show the density increasing at a steadily decreasing rate;

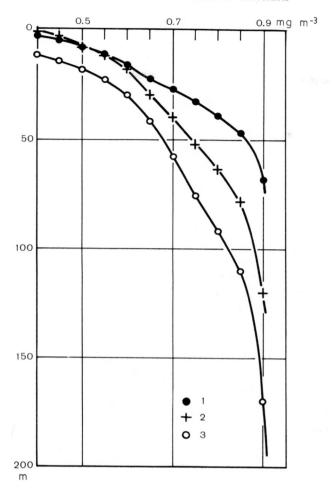

Fig. 4. Firn density versus depth at three locations on polar ice sheets. 1 – Ross ice shelf, from Langway (1975); 2 – Greenland, from Langway (1970); 3 – Vostok station, from Korotkevich (personal communication).

the three densification zones previously mentioned can often be identified by changes in slope (Figure 4).

At still greater depths in glaciers and ice sheets, small density changes occur due to the normal effect of temperature and pressure on the density of solids. The compressibility of ice is such that the density will increase in a temperate glacier by about 0.05% per 100 m of depth. In a polar glacier, however, a depth is reached at which the gradient of temperature increase with depth is great enough (>3° per 100 m) for thermal expansion to overcome the pressure effect. Thus, there is typically a density maximum produced. On the Ross Ice Shelf, for example, the depth to the density maximum is 150 m (Gow, 1963) [6]; at Byrd Station on the West Antarctic inland ice sheet, it is 1000 m (Gow, 1970) [8].

GRAIN SIZE

In the absence of dynamic metamorphism and paleoclimatic changes, grain size slowly increases by growth processes controlled by grain boundary energy. This is typical of the firn zone and of at least the upper half of ice sheets. The grain size increase proceeds rather uniformly in time, but the rate at which the grain size increases with depth (i.e., with time) increases with temperature. For example, at Byrd Station, a gradual increase in mean cross-sectional area from $5\,mm^2$ to $20\,mm^2$ occurs between depths of 100 and 300 m (Gow, 1963) [7], corresponding to a little more than 1000 years; on the Ross Ice Shelf the same increase occurs between 50 and 100 m, corresponding to about 250 years only.

Examination of deep cores that have penetrated through the depth corresponding to the boundary between Holocene and Pleistocene ice has shown an abrupt downward decrease in grain size at that boundary. Presumably the colder temperatures during the Pleistocene slowed the rate at which the grain size increases. Below this step, however, the grain size continues to increase.

Where the ice has undergone extensive plastic deformation other than that due simply to the densification process, major changes in the grain size can occur. Large grains are most commonly found, with dimensions as large as 100 mm, and with complex interlocking shapes. Fine-grained ice composed of regularly shaped grains of 1 mm size is also found, but intermediate grain size is not common. The association of ice of different texture and bubble content is typically in a layer-like pattern referred to as foliation. In the Byrd Station drill hole, the grain size increased abruptly about 350 m above the bottom, presumably marking the boundary of a basal zone that has undergone extensive deformation.

BUBBLES

At the firn–ice boundary, where the bubbles first become isolated, they have a tubular shape. As their compression increases, they slowly diminish in volume and assume a spherical shape. Air pressure within the bubbles is comparable to the glaciostatic pressure of the overlying ice. The total volume of air trapped in the ice is dependent upon the air density at the glacier surface, and thus on the surface elevation. Study of the total gas content as a function of depth can, therefore, lead to very valuable information on past elevations of a glacier surface. This has been particularly useful on the Antarctic and Greenland ice sheets.

CRYSTAL ORIENTATION

In the upper part of the firn layer in a dry snow zone, the ice crystals are oriented randomly, i.e., there is a random distribution of directions of the c-axes in space. As depth increases, a weak preferred orientation in the vertical direction occurs, Deeper in glaciers and ice sheets the fabric, i.e., the pattern formed by the c-axes, is very complex. The fine-grained ice included in the foliation of temperate glaciers shows a single maximum perpendicular to the plane of foliation. In marked contrast, the coarse-grained ice shows concentrations of c-axes in more than one direction, generally at similar angles to the foliation plane but distributed more or less symmetrically around the perpendicular to the plane. Patterns of 2, 3, and 4 maxima have been observed, as has a more or less even distribution throughout a 'girdle' around the vertical direction.

In the polar ice sheets, the fabrics may again be very complex. In the deep drill hole at Byrd Station, vertical concentrations become more and more intense, with an extremely strong vertical maximum developed at depths between 1200 and 1600 m (Gow, 1970) [8]. At greater depths the pattern becomes weaker again and, at some points within the ice sheet, two-pole maxima are found. In the Ross Ice Shelf highly preferred orientations are again found, with multiple pole patterns dominating (Gow, 1963) [7].

Because of the paucity of deep drill holes, much of the evidence concerning crystal orientation deep in ice sheets has come from geophysical measurements. Seismic wave velocities in a single crystal show variations of as much as 10%, depending on the direction of propagation. This translates to easily measurable variations in wave speed in anisotropic ice in glaciers. An extensive series of measurements on the West Antarctic inland ice yielded many indications of anisotropy, in some places at a depth of as little as 200 m (Bentley, 1971) [5]. In some places, observed seismic wave velocities could not be satisfied by any model in which the fabric consisted of a vertical maximum. Seismic reflections from the subglacial rock revealed that the basal ice in many locations must also be anisotropic with a non-vertical preferred c-axes orientation.

It is of great importance in dynamic glaciology to know what the fabrics are, particularly in the basal parts of glaciers and ice sheets where the stresses are the greatest, because ice crystals deform much more rapidly by displacements parallel to the basal plane then in any other direction. Thus the flow law that is appropriate to glacier flow is critically dependent on the ice fabric.

GEOCHEMICAL PROPERTIES

Glacial geochemistry is a very important and broad topic which we can only touch on here. The chemical characteristics of a glacier can be grouped into three categories: the stable isotopic composition of the ice itself, the occurrence of radioactive impurities in the ice (both man-made and naturally occurring), and the concentration of ordinary chemical impurities.

Studies of both deuterium and ^{18}O reveal essentially the same information – the temperature of a glacier's surface at the place and time at which the ice particle fell on that surface. In a dry snow zone the regular variation of age with depth yields an invaluable paleoclimatic record. Using total gas volumes to determine past surface elevations, as described above, the effect of elevation changes can be removed from the isotope record, leaving an indication of the climatically caused temperature changes.

Radioactive isotopes are chiefly of interest in determining the age of various layers within the ice. Identification of layers in which fallout from specific atmospheric thermonuclear bomb tests is found has proven to be an invaluable means of determining snow accumulation rates averaged over the past quarter of a century. Other radioactive isotopes can be used in the normal way to determine ages at greater depths. Unfortunately, no method has yet been found that is capable of dating the ice in the deeper parts of the polar ice sheets. There is a strong hope, however, that ^{14}C dating, which has heretofore been limited in its usefulness by the extremely large sample required, can be used in the future for ages as old as 100 000 BP using accelerator techniques that have drastically reduced the necessary sample size.

In terms of general chemical impurities, glacier ice can be no more pure than the precipitation from which it is formed. Other sources of impurities can be substantial, such as dust blowing from distant land, the direct avalanching of debris from surrounding rock slopes, or the incorporation of moraine from the glacier bed. Causes of layering within the ice are considered in Section 2.2. The impurities that fall with precipitation reflect, primarily, the presence of sea salts that have been picked up by air masses as they traversed the ocean. The record of chemical impurity with depth, therefore, can often reveal seasonal and longer-term changes in the distance to open ocean. This also can be useful as a paleoclimatic indicator. These impurities are of great importance for radioglaciology, since some of the electrical properties of ice depend very strongly on the degree of impurity (see Section 3.5 below).

2.2. Stratification and Temperature Regime of Glaciers

MAJOR GLACIER LAYERS

Previous chapters dealt rather briefly with a description of the stratified structure of glaciers, with emphasis on the physical aspects of their formation.

The major part of glaciers consists of ice formed from solid precipitation, i.e., snow. The formation of the two main layers, snow/firn and glacial ice, respectively, is the result of long-term and complicated metamorphic processes of snow–ice transformation. In marginal areas of the Arctic, Greenland, and Antarctica, rain plays a significant role. Rainwater penetrates into a layer of snow/firn and freezes, forming *ice lenses* – intermediate ice layers with a density higher than that of snow or firn. However, ice lenses formed due to rain have insignificant thickness and limited spatial extent. A firn layer covers the entire surface of glaciers over the whole year, except for short seasonal intervals. Its thickness depends on physical and geographical conditions – e.g., the snow/firn layer of the Arctic glaciers varies from one meter to several tens of meters thick. On the coast of Antarctica the snow/firn layer may not occur in summer, as has been mentioned earlier. In the central part of the continent it is from 100 to 150 m thick and from one hundred to several hundreds of years old (Zotikov, 1977) [1]. The term 'snow/firn layer' implies that this important surface layer consists of two layers that differ in structure and physical properties. Their major physical differences are in density and crystal structure. Thus, the density of Antarctic snow, depending on the geographical position of the area, varies from 0.30 to 0.42 g cm^{-3}. The density of the firn is higher – it varies from 0.60 to 0.82 g cm^3, depending on the depth below the surface (Kotlyakov, 1961; 1968) [2, 3].

Long-term studies have shown that the snow/firn layer comprises thinner layers. At the end of the 19th century Wilkes found rather regularly alternating layers in the cross-sections of icebergs. He suggested that blue bands in the snow layers could be a result of summer melting processes. Drygalski, a scientist from the German Antarctic Expedition suggested that the stratification of icebergs could be attributed to snow accumulation. While digging pits during the British National Antarctic Expedition, 1901–1904, R. Scott found that the snow consisted of a number of layers, separated at different depths by ice crusts. The annual layers of the Antarctic snow and firn were identified clearly by V. Shytt in 1958. Recently-developed methods make it possible to chart the development of the snow/firn layer. These methods are described in monographs by V.M. Kotlyakov (Kotlyakov, 1961; 1968) [2, 3].

Glacial ice is the main and the thickest layer. In Antarctica its thickness is about 4000 m. The layer consists of many individual ice layers, with boundaries marked by density anomalies. These layers were first identified by radar sounding in Antarctica and Greenland. The layer adjacent to the bedrock of a glacier is very important. That layer is characterized by a relatively high temperature and, therefore, by considerable plasticity. An intermediate layer between the glacial ice and bedrock plays the role of a very viscous lubricant that contributes to a glacier's movement. This intermediate layer is the most heterogeneous layer in the glacier. During the glacier's movement this layer traps bedrock fragments, sand, and other morainal materials. Sometimes the products of rock destruction involved in this movement form separate extended layers. In the process of movement these layers may be lifted from the bottom to the surface, thus forming clear inner reflecting surfaces. Layers of volcanic ash carried by dust storms from other continents were found in Greenland and, especially, Antarctica. Another physical phenomenon may also cause the formation of ice layers with different density, crystalline structure, and colour. This phenomenon, an extremely rapid, even catastrophic movement of some glaciers, appears to be due to variations in a number of geophysical and hydrometeorological conditions. The glaciers exhibiting such behavior are called surging glaciers. Surging glaciers are usually found in mountains, but they may also occur in the Antarctic ice sheet. Due to the rapid motion, layers within the surging glaciers are shifted relative to one another, and the consequent stratified system is maintained until the glacier discharges into the ocean.

On ice shelves (which are afloat) there occur, in addition to the shallow layers in the snow/firn ice mass, layers of salt ice at the interface between ice and seawater. Thus, ice shelves represent a complicated stratified system that is of particular interest to glaciology, especially radioglaciology, and to physical oceanography.

Even a brief review of the subject shows that a detailed study of stratification may answer many questions actively pursued by glaciologists, geophysicists, and geographers. It is quite evident at present that radar techniques are promising for studies of glaciers and composite multi-layered systems. Inhomogeneities in density and structure, moraine inclusions, and other features that, create internal reflections, have different impedance contrasts and therefore can be distinguished by radar techniques.

CHAPTER 2

TEMPERATURE DISTRIBUTION AND THE THERMAL REGIME OF GLACIERS

Ice is strongly influenced by thermal energy. It disturbs every atomic and molecular bond in ice and generates phase transformation: the solid matter is transformed into the liquid phase, i.e., water. It is evident that the physical properties of ice depend upon its temperature. Temperature fluctuations in ice may cause considerable changes in electrical conduction, electromagnetic wave absorption, elasticity, seismic wave velocity, etc. From this one can understand why there is an interest shown in detailed studies of the temperature and thermal regime of glaciers.

The temperature of each glacier is controlled by several factors. The upper and lower layers of the glacier have constant heat sources. The temperature of most of the inner part of the glacier is determined by heat conduction, water penetration, and ice advection. The upper layer, usually called the 'active layer', is a snow/firn layer characterized by annual temperature fluctuations. It is necessary to mention here that most observations are carried out in the active layer or in the next layer beneath it. This fact can be explained by the difficulties of drilling in glaciers. In theoretical studies use has been made of simplified models. In this section we try to discuss some problems of temperature measurements in different glaciers.

Generally, the active layer of a glacier is about 15 m thick. In winter, the ice is warmer than the air, so the heat flux is from the deeper layers to the surface of the glacier. From the equation of thermal conductivity it follows that a sinusoidal wave of cold with an amplitude of $-10°C$ causes a minimum temperature at the depth of 10 m in 5.5 months. Due to this fact it is then impossible to determine seasonal variations of temperature at depths below 15 to 20 m. Field measurements are in agreement with this conclusion. In the Antarctic ice sheet near Wilkes station the temperature is equal to $0.85°C$ at a depth of 10 m, and to $-0.2°C$ at a depth of 16 m. On the Axel Heiberg ice cap (Northern Canada) seasonal variations of temperature could be detected down to 11 m, and down to 16 m on top of one of the domes in the region of valley-glacier ablation.

In most regions the propagation of heat energy in the active layer in summer differs from that in winter. In summer, melting at the surface can be observed almost everywhere (with the exception of the inner regions of Antarctica and Greenland). Freezing of percolating water releases a sufficient amount of heat to cause a temperature increase in the upper layers of the glacier. Sverdrup showed experimentally that at a depth of 3 m on the Isaksen glacial plateau (Spitzbergen) the winter cold wave reached its minimum on the 25th of June, when the temperature at a depth of 10 m was $-0.2°C$.

It is worth mentioning here that the temperature in the ablation zone is higher than it might be in the absence of a snow cover. Quite an opposite situation is observed in the accumulation zone where the snow cover is a part of the glacier and plays the role of a thermal insulator. Table IV shows that air and firn temperatures coincide at shallow depths in the active layer. Several discrepancies in values can be explained by an insufficient number of air temperature measurements.

Another interesting comparison of temperatures measured at different glacier points is given in Table V. It is well known that air temperature decreases with

TABLE IV
Comparison of firn and mean annual air temperatures in dry snow zones

Site	Latitude	Longitude	Firn temperature (°C)	Depth (m)	Air temperature (°C)	Author
Northice	78°04 N	80°29 W	−28	15.0	−30	Bull
Camp Century	77°10	61°08	−24	10.0	−23.6	Weertman
Site 2	76°59	56°04	−24	8.0	−24.4	Diamond
Eismitte	71°11	39°56	−29	6.6	−30	Sorge
Byrd Station	80°00 S	120°00	−28	15.0	−29	Gow
South Pole	90°00	–	−51	10.0	−51	Sharp

TABLE V
Temperature at different points in glaciers

Glacier	Region	Temperature (°C)	Depth (m)	Author
White Glacier (Axel Heiberg, Canada)	Firn accumulation Along the equilibrium line Ablation	−9.5 −16.0 −13.0	8 8 8	Müller
Jackson Dome (Franz-Josef Land, USSR)	Firn accumulation Ice accumulation Ablation	−3.0 −9.0 from −6.0 to −10.5	20 20 20	Krenke
Vestfonne Dome (North-eastern Spitzbergen)	Accumulation Ablation	−3.0 from −7.0 to −10.0	14	Schytt

height. Thus, it should be expected that the temperature in the accumulation zone will be lower than that in the ablation zone. The table shows that this rule is not always true. This deviation can be attributed to refreezing of melt water in the firn.

TEMPERATURE DISTRIBUTION WITH DEPTH IN THE GREENLAND ICE SHEET

A hole drilled in the Greenland ice sheet in 1966 has turned out to be very useful for a further study of englacial temperatures. The hole was drilled at Camp Century (77°10′ N, 61°08′ W), where the ice is 1387 m thick. Figure 5 gives values of measured temperatures plotted by Hansen. The plot shows a decrease of temperature in the upper layer (10–154 m) from −24°C to −24.6°C. After passing the minimum the ice temperature increases gradually until it reaches −13°C at the bed. The temperature gradient in the lower layer (3 m) is constant, formed by the geothermal flux.

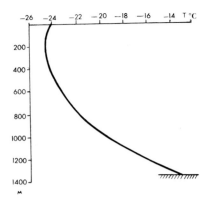

Fig. 5. Temperature in the Camp Century Hole.

TEMPERATURE DISTRIBUTION WITH DEPTH IN THE ANTARCTIC ICE SHEET

Recent field observations of the temperature regime of the Antarctic ice sheet were carried out in the marginal and interior regions of Antarctica. In the marginal regions the temperature was measured in shallow holes, drilled down to 50 m. The measurements show a decrease in ice temperature with depth at a distance of 500 km from the coast. Figure 6 (Kartashov's data) shows temperature variations with depth (from the surface down to 70 m). The measurements were carried out by S.N. Kartashov on the route from Mirny Observatory to Pionerskaya station. It is seen that the temperature gradient decreases with distance from the edge of the ice sheet. A hole drilled near the ice edge where the ice velocity is very small ($0.3\,\text{m}\,\text{y}^{-1}$) demonstrates another pattern. Temperature measurements in the upper layer of the immobile ice down to 75 m show the dependence of the tempera-

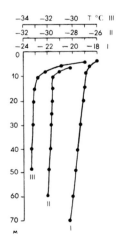

Fig. 6. Temperatures in the upper part of the ice in East Antarctica (after Kartashov), changing with the distance from the ice edge: I – 42.5 km distance; II – 143 km distance; III – 216 km distance.

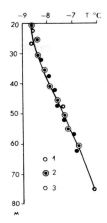

Fig. 7. Temperature in the stagnant ice in the vicinity of Mirny Observatory: 1 – Zotikov's measurements (1959); 2 – V.N. Bogoslovsky's measurements (1957); 3 – Kapitsa's observations (1956).

ture on depth. The results of the measurements are given in Figure 7. Figure 7 shows that the increase in ice temperature with depth is close to linear. A similar relationship was obtained by Cameron during his measurements in holes in immobile ice in the vicinity of Wilkes station (Zotikov, 1977) [1]. Data on surface temperatures obtained in different areas on the Antarctic ice sheet fail to give information sufficient for definite conclusions. According to available data, the temperature gradient in the layer between 30 and 50 m in the central areas of the East Antarctic glaciers is negative. The conclusion was made on the basis of

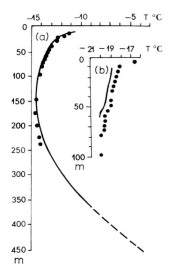

Fig. 8. Ice temperatures in the hole near Mirny Observatory: (a) 5 km from the ice edge; (b) 50 km from the ice edge. Dots – Zotikov's measurements; solid line – Bogoslovsky's data; dashed line – extrapolated values.

measurements carried out at Pionerskaya, Vostok, Komsomolskaya, Vostok-1, Sovyetskaya, and the Pole of Relative Inaccessibility. Reverse or negative gradients obtained from these measurements appeared to be large (5.1°C per 100 m at Pionerskaya). Measurements made by A.V. Krasnushkin during the Vostok-South Pole traverse fail to give any evidence of temperature gradients (Zotikov, 1977) [1].

The results of temperature measurements in deep holes are important, as they throw further light on the problem of the thermal regime of the Antarctic ice sheet. The first hole of that kind (300 m deep) was drilled at Byrd station in the central part of West Antarctica, in 1957. The second, 355 m deep was drilled near the Mirny. A curve showing the temperature change with depth at Mirny is given in Figure 8. As is seen from the figure, there is a pronounced reverse temperature gradient. The maximum gradient is at the surface. It decreases down to 150–170 m depths, where it changes sign. Below that the temperature increases steadily. Figure 9 gives the values of temperature in the ice near Byrd station. The comparison of the two plots (Figures 8, 9) indicates the differences in these two areas. The measurements of temperature carried out at Vostok are of particular interest.

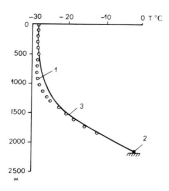

Fig. 9. Ice temperature, Byrd Station: 1 – measurements by Gow (1967); 2 – pressure-melting point for ice at the bedrock; 3 – temperature curve calculated for the conditions at the station.

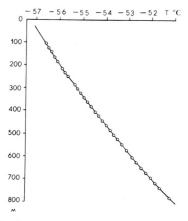

Fig. 10. Ice sheet temperatures at Vostok Station.

In a hole about 900 m deep the measured temperature varies nearly linearly with depth (Figure 10). Temperature in an ice shelf was measured in a 255 m hole drilled to the base of the Ross Ice Shelf. The mean surface temperature of the Ross ice shelf obtained from multiyear observations was about −22.3°C. In the ice, there is an increase in temperature with depth that is at first slight, and then becomes more pronounced beyond 100 m depth. The ice temperature increases to −4.4°C at the depth of 250 m. In the layer adjoining the sea water of 33‰ salinity it is −1.8C.

Recently, a method has been developed, based on the absorption of radar signals, for determining a vertically-averaged effective temperature in glaciers. Direct measurement of glacier temperatures combined with radar sounding has turned out to be very promising for the understanding of the thermal regime of glaciers, which is one of the most important problems in the field of glaciology.

2.3. Glacier Regimen

All glaciers, regardless of type, move, grow, and shrink according to one basic process and its variations: the fall of snow on the surface, the transformation of that snow to ice, the movement of ice downward and outward, and the loss of the ice at, or near, the ice margin by melting or the calving of icebergs. Where it is not melted away, the snow that is deposited on the surface is buried by successive snow falls. As time passes, snow continues to accumulate on the surface, and the deeper layers are gradually compressed by the weight of the overlying layers until they are transformed into solid ice.

The transformation from snow into ice that we have described takes place only on that part of the glacier where the annual accumulation of snow exceeds the annual melt, called the accumulation zone. Where the melt rate, averaged over a year, exceeds the accumulation rate, solid ice will be found right at the surface; this is the ablation zone.

The mass balance of a glacier, i.e., the net change in mass over some period of time, depends upon the balance between the total accumulation and total ablation (mass loss). In temperate and subpolar glaciers, particularly those that do not terminate in a body of water, the principal ablation loss is by melting at the surface. In polar ice sheets, or glaciers ending in deep water, the principal loss usually is calving of icebergs.

GLACIER MOVEMENT

The movement of a glacier occurs in two different ways: by deformation of the ice within the glacier and by sliding of the glacier over its bed. Generally speaking, sliding is likely to predominate if the base of the ice is at the pressure melting point, so that a lubricating film or layer of water is present. In fast-moving glaciers, particularly outlet glaciers and ice streams, it is virtually certain that sliding is the dominant process. On the other hand, when the ice is frozen to its bed sliding will be largely or completely eliminated, and deformation will become the dominant process. Because the deforming stresses increase with depth in the ice, because the deformational response, or creep, of the ice is non-linear – increasing at a greater than linear rate as the stress increases (see below) – and because creep is enhanced by the warmer temperatures near the base of the ice, plastic deformation also is concentrated near the base of the glacier or ice sheet.

A glacier is driven by the force of gravity acting down the surface slope, so the horizontal component of movement is in that direction. In a simple glacier, the component of ice particle movement normal to the surface will be downwards in the accumulation zone and upwards in the ablation zone (Figure 11). In the central part of ice sheets and high in the accumulation zone on valley glaciers, the horizontal movement is small, and the principal particle motion is downward. As one moves down a flowline, however, the mass flux increases as the area of accumulation at the surface upstream increases. Thus if the glacier thickness and width

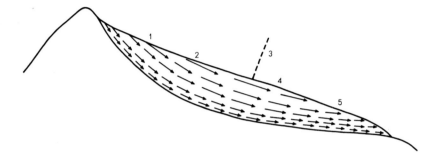

Fig. 11. Longitudinal section of an idealized glacier, with velocity vectors. Note changes in velocity both vertically and horizontally. 1 – accumulation zone; 2 – extending flow; 3 – equilibrium line; 4 – compressing flow; 5 – ablation zone.

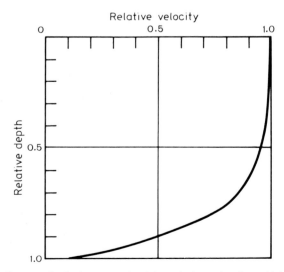

Fig. 12. Schematic diagram of velocity versus depth in a glacier or ice sheet. Velocity and depth scales are normalized.

change slowly, the horizontal velocity in the accumulation zone will increase downstream. Generally, the distance downstream beyond which the horizontal velocity exceeds the vertical velocity is not very great. Since both deformational and sliding movement is concentrated in the bottom of the ice, the horizontal velocity changes very slowly with depth near the surface. Thus throughout a major part of the ice thickness it is true to a good approximation that vertical columns remain vertical as they move downstream. A representative velocity vs. depth curve is shown in Figure 12. For a polar ice sheet in which there is a large temperature increase with depth, the upper zone in which the horizontal velocity changes little with depth may extend proportionally much closer to the bedrock.

DEFORMATIONAL FLOW

A great deal of experimental and theoretical work has been devoted to attaining a better understanding of the so-called 'flow law' of ice, i.e., the way in which ice responds to a deforming stress. As a generalized summary we can say that the strain rate in shear of the ice is proportional to the shear stress raised to a power that is commonly (but not universally) taken to be about 3. The constant of proportionaliy increases exponentially with temperature.

The basal shear stress, τ_b, can be expressed to a good accuracy by the simple expression $\tau_b = \rho g h \sin \alpha$, where ρ is the density, g the acceleration due to gravity, h the ice thickness and α the surface slope. It is an experimental fact that in most places the shear stress at the bed is between a few tenths of a bar and 1 bar. This means that a fairly consistent inverse proportionality exists between the ice thickness and the surface slope. That inverse proportionality is particularly well observed where the bedrock topography is small or where averages over distance many times the ice thickness are taken.

The flow law of ice and the relationship between surface slope and ice thickness can be used to develop a theoretical equilibrium profile of an ice sheet. Such a profile is generally parabolic in shape and agrees quite nicely with the observations in many parts of Greenland and Antarctica. In other places both on the ice sheets and on other ice caps, the agreement is not so good, indicating perhaps the insufficiency of the theory or perhaps systems that are not in steady state.

When one takes the longitudinal stress parallel to the direction of flow into account, it is found that two states of flow are possible: one, called extending flow, in which the longitudinal stress exceeds the transverse stress (tensile stresses are taken as positive), and the other, in which the transverse stress is greater (compressing flow). In extending flow the velocity vectors are downward relative to the surface and the velocity increases downstream, whereas for compressing flow the vertical component of the velocity is toward the surface and velocities decrease downstream. To a first approximation, extending flow is associated with the accumulation zone and compressing flow with the ablation zone on a glacier, although an important role is also played by the bedrock topography. In extending flow the longitudinal stress is actually tensile near the surface, conducive to the formation of transverse crevasses if the stress is great enough.

GLACIER SLIDING

Two mechanisms occur at the base of a glacier that allow the glacier to slide on its bed: pressure melting and enhanced creep. The resistence of obstacles on the bed to the movement of the glacier increases the pressure on the upstream sides of those obstacles, causing the ice to melt. The melt water flows around the obstacle and re-freezes in the lower-stress region on the downstream side. To maintain this process, the latent heat of fusion must be transferred from the downstream to the upstream side of the obstacle, by conduction through the ice and the bedrock. Because of this heat conduction requirement, the pressure-melting process becomes less efficient as the size of the obstacles increase. Furthermore, it will not work unless the temperature of the ice is at the melting point.

Enhanced creep also stems from the excess stress on the upstream side of an obstacle on the bed. The excess stress will cause an accelerated strain rate, in accordance wth the flow law of ice, that will be maintained for a distance comparable to the size of the obstacle. Thus the velocity enhancement (the integral of the strain rate enhancement), increases as the size of the obstacle increases, i.e., it becomes more efficient as the obstacle size increases, in contrast to the pressure-melting mechanism.

It follows that if only the pressure melting process were effective, a glacier could not slide on any bed that had large obstacles. Conversely, if only the enhanced plastic flow process were present, the glacier could not slide if there were any small obstacles. Although much remains to be learned about the sliding process, particularly in terms of a quantitative description, it seems virtually certain that a combination of these two processes is required, thus allowing the glacier to slide around obstacles both large and small. We note that only the plastic deformation process can operate at temperatures below the melting point, so that presumably a cold-bedded glacier can slide only if the small-scale roughness is extremely low.

It is intuitively obvious that the presence of water at a bed will enhance the sliding process. Obstacles will be drowned; if subglacial water pressures are high enough, they can even tend to lift a glacier off its bed. Nevertheless, a quantitative explanation of the action of water at the bed of a glacier still remains to be developed. One particular phenomenon probably associated in some way with water at the basal ice rock interface is that of surging. Typically, the ice, after moving slowly and quietly for many years, starts to move much more rapidly, maintaining for a short period of time a velocity an order of magnitude or more greater than its normal velocity. Such a surge transfers a large mass of ice from the upper part of a glacier to the lower part. The terminus may advance suddenly as a result, but this does not necessarily happen. The surge ends and the glacier returns to another long quiescent stage before the rapid advance is repeated. Although most glaciers show no evidence of ever surging, the surge phenomenon is certainly not uncommon. The question then arises as to whether surges are possible in ice caps and even in the major ice sheets. Some studies based on computer modeling and on examinations of Antarctic surface topographic profiles suggest that surges not only can occur in the Antarctic ice sheet but have occurred. On the other hand, other investigators read the evidence in quite a different way. A primary type of observational evidence comes from radar studies of internal layers (see Sections 2.1 and 6.8).

References: Chapter 2

[1] Zotikov, I.A. *Thermal regime of the Antarctic ice cover.* Leningrad, Hydrometeoizdat, 1977, 168 pp.
[2] Kotlyakov, V.M. 'Snow cover of Antarctica and its role in the present glaciation of the continent. *J. Glaciology*, 1961, No. 7, 240 pp.
[3] Kotlyakov, V.M. *Snow cover of the Earth and glaciers.* Leningrad, Hydrometeoizdat, 1968, 479 pp.
[4] Paterson, W.S.B. The Physics of Glaciers. 2nd Edn. Pergamon 1981, 380 pp.
[5] Bentley, C.R. 'Seismic anisotropy in the West Antarctic ice sheet'. In: *Antarctic Snow and Ice Studies II.* American Geophys. Union, Antarctic Research Ser., Vol. 16. Washington, 1971, pp. 131–177.
[6] Gow, A.J. 'Results of measurements in the 309-meter bore hole at Byrd Station, Antarctica'. *J. Glaciol.*, 1963, **4**, 771–784.
[7] Gow, A.J. 'The inner structure of the Ross Ice Shelf at Little America V, Antarctica, as revealed by deep core drilling'. *General Assembly of Berkeley, IASH Pub.* No. 61, 1963, pp. 272–284.
[8] Gow, A.J. 'Preliminary results of studies of ice cores from the 2164 m deep drill hole, Byrd Station, Antarctica'. *International Symp. on Antarctic Glaciol. Exploration (ISAGE). IASH Pub.* No. 86, 1970, pp. 78–90.
[9] Langway, C.C. 'Stratigraphic analysis of a deep ice core from Greenland'. *Geol. Soc. Amer. Special Paper 125*, 1970, 186 pp.
[10] Langway, C.C. 'Antarctic ice core studies'. *Antarctic J. U.S.*, 1975, **10**, 152–153.
[11] Post, A., LaChapelle E.R. *Glacier ice.* Seattle, Univ. of Washington Press, 1971.

CHAPTER 3

ELECTROMAGNETIC WAVE PROPAGATION IN ICE

3.1. Crystalline Structure of Ice

A review of the physical properties of ice naturally starts with a description of the water molecule. The water molecule is a combination of two hydrogen atoms and one oxygen atom, and is triangular in shape. The fact that the molecule is not linear has been know for many years both from studies of the specific heat of water vapor and from the fact that water is a polar molecule with a permanent electric dipole moment. Furthermore, the triangle must be isosceles because an asymmetrical molecule would be unstable. The explanation for the triangular shape appears when the electronic structure of the oxygen atom is examined. The eight electrons surrounding an oxygen nucleus are found two each in two spherical shells ($1s$, and $2s$), the other four being found distributed among the three dumbbell-shaped shells ($2p$) that represent the next higher energy level. There are thus two vacancies, one in each of two of the dumbbells. The water molecule is formed by a homopolar (covalent) bond between two hydrogen atoms with their single electrons and the two holes around the oxygen atom. The repulsion between the positive charges on the hydrogen atoms forces the angle between the legs of the triangle open beyond $90°$ to a value of about $104.5°$. The distance between the oxygen and hydrogen nuclei is slightly less than 10^{-4} microns.

The sharing of its electron with the oxygen atom leaves a net positive charge in the vicinity of the hydrogen nucleus, and results also in a distortion of the electron clouds around the oxygen atom in such a way as to produce negative charge concentrations about two points on the side opposite the hydrogen atoms, separated from each other by about the same angle, but lying in a plane perpendicular to the HOH plane. There is then an electrostatic attraction that can form between a proton in one water molecule and the negative charge on another. At temperatures below the boiling point, this attraction is enough to link molecules together in a hydrogen bond. The liquid phase of water is characterized by a process of rapid formation and breakage of hydrogen bonds. When thermal motion is decreased enough by cooling, the hydrogen bonds pull the molecule together into the rigid crystalline structure of ice.

At normal pressures, the crystalline structure corresponding to the lowest energy state involves an increase in the angle of the apex of the water molecules from $105.5°$ to $109.5°$, appropriate to a tetragonal (four-fold) coordination between each oxygen atom and its nearest neighbors. This, in turn, leads to sheets of crinkled hexagons stacked one on top of another; this characterize the structure of ice (see Figure 13). The hexagonal crystal exhibits cylindrical symmetry, with physical

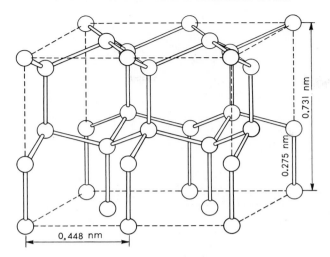

Fig. 13. Expanded view of oxygen atom positions in the ice structure. Dimensions shown in this and later figures are those for a temperature of 77°K (Kamb, 1968). (From *Structural Chemistry and Molecular Biology*, ed. Alexander Rich and Norman Davidson. San Francisco: Freeman & Co., Copyright © 1968.)

properties that generally are a function of angle relative to the c-axis, the direction normal to the stacked sheets or basal planes. The distance between closest oxygen atoms is 2.76×10^{-4} microns (2.76 Å), both in the basal planes and parallel to the c-axis.

The hydrogen atoms can not be located by X-rays because of their small size. Studies of the infrared and Raman spectra of ice showed that the molecular vibrations were similar to those of a free water molecule, leading to the so-called Bernal and Fowler rules: (1) there is one and only one proton per O–O linkage; (2) there are two, and only two, protons associated with each oxygen atom, i.e., the molecular identity of the water molecule is maintained in the structure of ice. These rules are not enough to fix all proton positions; proton disorder is indicated by the relatively high residual entropy of ice at 0 K. Not until the mid-1950s did direct experimental evidence from neutron diffraction experiments (Peterson and Levy, 1957) [11] confirm both the disordered structure and the location of the protons at a distance of 10^4 microns (1.0 Å) from the centers of the oxygen atoms (Figure 14). A photograph of an ice crystal model is shown in Figure 15.

As with many other substances, the electrical properties of ice, with which we are concerned in this monograph, stem not from the basic crystalline structure but from violations of that structure; namely, from point defects. These defects are of three kinds: (1) violation of tetrahedral bonding, i.e., the occurrence of vacancies or interstitial molecules; (2) violation of the first Bernal/Fowler rule, i.e., the occurrence of either two protons or none along a O–O bond, producing an orientational (Bjerrum) defect; and (3) violation of the second Bernal/Fowler rule, i.e., the occurrence of either one or three protons associated with an individual oxygen atom, i.e., an ionic defect.

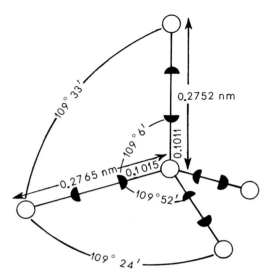

Fig. 14. Average configuration of a molecule in D_2O ice at $-50°$ as determined by neutron diffraction (after Peterson and Levy, 1957). The deuterons appear in symmetrical pairs of half-deuteron positions.

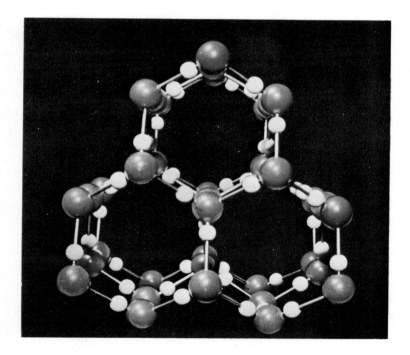

Fig. 15. Photograph of a model of an ice crystal, looking parallel to the c-axis. Large balls represent oxygen atoms; smaller balls represent hydrogen atoms.

Consider first the ionic defects. Ionic states are most simply produced by a proton jump from one end of an O–O bond to the other. Because of its higher energy level relative to the non-ionized state, this initial ionic state will, at best, be metastable. However, the finite time required to jump the intervening energy barrier will, in a small fraction of the cases, allow the proton to remain in its displaced position long enough for another proton from the same molecule in turn to jump to a third oxygen, thus separating the ionic states so that recombination is not directly possible. The two ions thus produced are then free to move through the crystal until they accidentally encounter ions of the opposite sign and are destroyed.

Because the generation of an ion state is thermally activated, the concentration will be proportional to an Arrhenius function, i.e., a function of the form $\exp(-h/kT)$, where h is the activation energy (enthalpy) of the formation process, usually a fraction of an eV, and K is Boltzman's constant. A similar expression applies to the movement of a proton. Electrical conduction in ice depends centrally upon the movement of ion states through the crystal, so one would expect the electrical conductivity to increase exponentially with temperature, a fact that is experimentally supported.

For radioglaciology, the most important defects are the orientational defects. These are also produced in pairs by the jump of a proton, in this case from its position as the single occupant of an O–O bond to another bond linking the same molecule, the new bond already being occupied by a proton at its other end. Alternatively, the orientational defect pair can be viewed as produced by the rotation of a molecule around one of its bonds through an angle of 120°. Again the defect pair can be stabilized by a further separation through another proton jump (or another molecular rotation). It is apparent that this process also is thermally activated; the activation energies are about 0.7 eV for the generation of orientational defects and about 0.25 eV for diffusion of the defects through the crystal. At $-10°C$ orientational defects occur at about 1 molecule in 10^7, a much higher concentration than for ion states, whose relative proportion at $-10°C$ is only about 10^{-12}.

3.2. The Electrical Properties of Monocrystalline Ice

Consider now the response of an ice crystal to an externally imposed electrical field, E. The response will be of two types, polarization and conduction. Since the identities of the water molecules are largely maintained in the ice crystal structure, their polar characteristic is also maintained. Thus, there will be a tendency for the dipoles to line up parallel to the externally-imposed electric field. That polarization will take place principally through proton jumps associated with orientational defects. The externally applied field has made some of the energy minima associated with the protonic positions slightly deeper than others, but that change is very small compared with the height of the energy barrier that must itself be thermally overcome. Thus, there will be a time delay in the development of the polarization, the length of which will depend on temperature. This becomes important particularly when we consider the application of a time-varying electric field. Let the electric field $E = E_0 \exp(i\omega t)$. Then the electrical displacement $D = \varepsilon E$, where the permittivity $\varepsilon = \varepsilon_0(\varepsilon' - i\varepsilon'') = |\varepsilon|\exp(i\delta)$, with $\tan\delta = \dfrac{\varepsilon''}{\varepsilon'}$; ε_0 is the permittivity of free space, $\varepsilon' - i\varepsilon''$ is the complex relative permittivity, or dielectric constant, and $\tan\delta$ is known as the loss tangent. This leads to $D = D_0 \exp(i(\omega t - \delta))$, where $D_0 = |\varepsilon|E_0$. The polarization depends upon the rate at which the protons will jump to a preferential orientation, and is thus proportional to $\exp(-t/\tau)$, where the characteristic relaxation time τ depends on the probability of a protonic jump, and hence exhibits a temperature dependence represented by an Arrhenius function. This characterization then leads to the standard Debye equations:

$$\frac{\varepsilon}{\varepsilon_0} = \varepsilon_\infty + \frac{\Delta\varepsilon}{1 + i\omega t}; \qquad \Delta\varepsilon \equiv \varepsilon_s - \varepsilon_\infty,$$

whence

$$\varepsilon' = \varepsilon_\infty + \frac{\Delta\varepsilon}{1 + \omega^2\tau^2}; \qquad \varepsilon'' = \frac{\Delta\varepsilon\omega\tau}{1 + \omega^2\tau^2};$$

$$\tan\delta = \frac{\Delta\varepsilon\omega\tau}{\varepsilon_{s'} + \varepsilon_\infty\omega^2\tau^2},$$

where ε_s and ε_∞ are the static and high-frequency values of ε', respectively, and $\Delta\varepsilon$ is known as the dispersion strength. A diagrammatic plot of a Debye spectrum is show in Figure 16. Note that ε'' (and consequently $\tan\delta$), which characterizes the energy absorption in the medium, attains a maximum when $\omega = \tau^{-1}$. The corresponding frequency, $f_c = (2\pi\tau)^{-1}$, is known as the characteristic frequency.

There are two other polarization phenomena whose contribution to the permittivity is small compared with the molecular rotation just considered, but that determine the value of ε_∞. These involve slight distortions of the molecular bonds and polarization of the molecular electron cloud. The former have characteristic frequencies in the infrared and the latter in the ultraviolet.

The conduction of a steady (d.c.) current through ice involves the diffusion of both ionic and orientational defects. The diffusion of either type of defect alone will lead quickly to a fully polarized state in which further proton transfer is impossible. However, the passage of an orientational defect followed by the passage of an ionic defect (or vice versa), results in the net transport of one proton and a restoration of the original configuration. The ratio of the current density to the applied electric field gives the d.c. conductivity: σ_s.

For an alternating current, or electromagnetic wave, it can be shown readily from Maxwell's equations that the dielectric conductivity $\sigma_{ac} = \omega\varepsilon_0\varepsilon''$. The dielectric conductivity includes the contribution of σ_s.

As seen from the Debye equations, ε'' (thus also σ_{ac}) will have a peak at the critical frequency. It is a rapidly varying function of both temperature and frequency for f near f_c. At higher frequencies, $\varepsilon'' \to 0$ as ω increases. On the other hand, it also follows from the Debye equations that σ_{ac}, which is proportional to

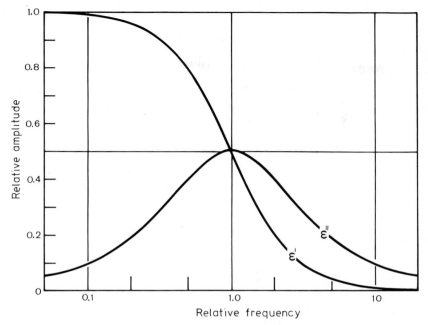

Fig. 16. Idealized Debye spectrum, normalized, showing both the real and the imaginary parts of the dielectric constant. Frequencies are normalized to $f_c = 1$, where f_c is the characteristic relaxation frequency.

$\omega \varepsilon''$, approaches a frequency-independent value that we designate σ_∞ by analogy with ε_∞: $\sigma_\infty = (\Delta \varepsilon / \tau) \varepsilon_0$. Of course σ_∞ is still essentially an exponential function of temperature through its dependence on τ, since $\Delta \varepsilon$ varies relatively slowly with τ. It may be useful to note that, at $f = f_c$, i.e., when $\omega = \tau^{-1}$, $\sigma_{ac} = \sigma_\infty/2$.

3.3. Values of the Electrical Parameters

Because of the polar nature of the H_2O molecule in ice, the static dielectric constant is very large. Anisotropy is well exhibited: actual values of ε_s are around 105 and 90 at 0°C, increasing to about 135 and 105 at −50°C, for E parallel and perpendicular to the crystal's c-axis, respectively (Humbel, et al., 1953) [7]. The effect of pressure is small; it has not been measured on single crystals, but amounts to an increase of only about 1% per kilobar for polycrystalline ice (Chan, et al., 1965) [2].

Across a broad frequency range (about 10^6–10^{13} Hz), $\varepsilon_\infty = 3.2$ at 0°C. ε_∞ is isotropic within at least 0.5% (Johari and Jones, 1978) [9]. In contrast with ε_s, ε_∞ shows only a weak temperature dependence – it decreases about 1% between 0° and −15°C (Johari and Charette, 1975) [8], and another 1% down to −75°C (Johari and Jones, 1978) [9]. The pressure dependence of ε_∞ does not appear to have been investigated. In the visible region all molecular absorption bands have been passed and the dielectric constant has simply its optical value of 1.72 due to electronic polarization alone. Above 10^{17} Hz, ε reaches its X-ray value, close to 1. The value of ε' as a function of frequency is summarized in Figure 17.

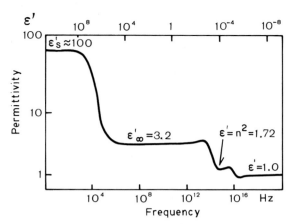

Fig. 17. Schematic representation of the behavior of the dielectric constant of ice, ε', as a function of frequency at a temperature near $-10°C$ (from Fletcher, 1970).

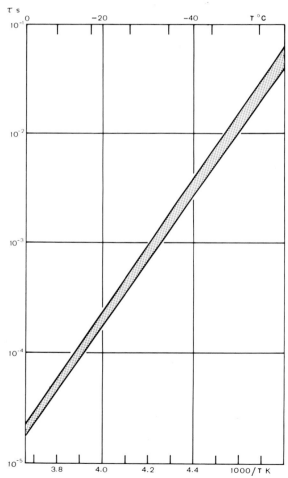

Fig. 18. Characteristic relaxation time as a function of absolute temperature for pure monocrystalline ice.

The characteristic relaxation time, τ, has a value of about 2×10^{-5} s near 0°C with an activation energy of about 0.58 eV (Humbel, et al., 1953; von Hippel, et al., 1971) [7, 15] (Figure 18). This means that $f_c \simeq 8$ kHz near 0°, decreasing to only 30 Hz at -50°C. The pressure effect amounts to an increase in the relaxation time of about 10% between 0 and 500 bars, independent of temperature (Ruepp, 1973) [13].

The measurement of σ_s in laboratory samples of ice has suffered from many experimental difficulties, and reported values vary over an order of magnitude. The lower values are probably more nearly correct for the true bulk d.c. conductivity, i.e., about $10^{-8} \Omega^{-1} \text{m}^{-1}$ at -10°C (Bullemer, et al., 1969) [1], decreasing by a factor of about 20 at -50°C (activation energy between 0.35 eV and 0.4 eV). Pressure increases σ_s by about 10% at 500 bars (Taubenberger, et al., 1973) [14].

σ_∞ at 0°C is about $4 \times 10^{-5} \Omega^{-1} \text{m}^{-1}$ when the electric field is parallel to the c-axis, and about 10% less with the field perpendicular to the c-axis (Ruepp, 1973) [13]. In accordance with the measurements of τ, the activation energy is slightly less than 0.6 eV.

3.4. Pure Polycrystalline Ice

Our concern in this section is with measurements taken on samples of polycrystalline ice prepared in a laboratory. Measurements taken on samples taken from the field will be discussed along with the other field results in Section 6.1. below.

Generally speaking, values of electrical parameters for pure polycrystalline ice are close to those for single-crystal ice; where anisotropy appears, they are close to values taken with the electric field perpendicular to the c-axis. Such minor differences as do occur in some experimental results are either outside the temperature, pressure, or frequency range of interest in field radioglaciology, or are too small to be observable in the light of experimental errors inherent in field measurements. For example, in Glen and Paren (1975) [4] listed values of ε'_∞ vary between 3.195 at 0°C and 3.158 at -60°C, in accordance with very careful and consistent measurements by Gough (1972) [5] (which had to be extrapolated from measurements at frequencies below 60 kHz). The value of ε_∞ at 0° is essentially identical to the single-crystal determination by Johari and Jones (1978) [9] discussed in Section 3.3, although the value at -60° for the single-crystal measurements was about 3.14. There are, however, two further aspects of pure polycrystalline ice that should be considered because of their potential importance in the field: the deformational and the metamorphic histories of the ice.

The influence of plastic deformation on the electrical properties of ice is a matter of substantal disagreement. Most of the evidence from laboratory ice suggests that a deformation dependence does occur but relates to frequencies well below f_c, and is therefore out of the range of primary interest to radioglaciology. However, at least one measurement on glacier ice has shown a dependence of τ on deformation (see Section 6.1).

The mode of formation of the ice and its temperature history are very important. Electrical properties vary significantly depending upon whether the ice is formed by direct freezing from the liquid state, by cold metamorphism of snow, or by metamorphism of snow followed by annealing at temperatures close to the melting point. It is probable that the several-orders-of-magnitude difference in τ_s exhibited by glacier ice of various types must be attributable, at least chiefly, to its metamorphic history, but very few laboratory measurements on this aspect exist, and results are still not definitive. Again, the effect is primarily noticeable at frequencies less than f_c; there is no evidence that the metamorphic history has any significant effect on the electrical properties at radar frequencies. However, the direct evidence is far too scanty to draw any firm conclusions.

3.5. Impure Polycrystalline Ice

The discussion so far has been related to pure ice. Such measurements reveal the basic physical processes involved in the electrical response of the ice, but are not directly relevant to ice found in glaciers and ice sheets which did not form directly from distilled water. It is very important that we consider the effect of impurities.

A great many laboratory experiments have been made on impure or 'doped' ice samples, but the vast majority of them have emphasized doping by substances that enter directly into the ice lattice, particularly ammonia, hydrochloric acid, ammonium hydroxide, and ammonium fluoride, because of the important information obtained thereby about the protonic processes in the ice. We, however, are more interested in the effect of the impurities, chiefly sea salts, found most commonly in precipitation. A good sampling of the types of results found, showing their complexity, is given by Gross et al. (1978) [6]. Considering an impurity level of the order of 10^{-6} molar, appropriate to the purest parts of the Antarctic ice sheet, we find typically that the values of τ and σ_{ac} are little changed near 0°, but that the activation energy is substantially diminished so as to cause an increasingly large difference as the temperature drops. For example, there is a negligible effect on τ at 0°, but a reduction by an order of magnitude relative to pure ice at $-50°C$. Note that, since $\sigma_\infty \sim \tau^{-1}$, a diminishment of τ corresponds to an increase in the absorption rate by a proportional amount: σ_∞ at $-50°C$ is an order of magnitude greater for ice with a 10^{-6} molar concentration of impurity than for pure ice. As the concentration of the impurity is increased, the effects are also increased, although not in a simple way. Fortunately, there is neither evidence to suggest, nor reason to believe, that ε_∞, whose value does not depend upon protonic defects, is affected by the concentrations of impurity found in glacier ice.

3.6. Snow

Since this book is concerned with the study of glaciers and ice sheets, our concern with snow is primarily to obtain sufficient information to make corrections for the snow and firn that overlie solid ice.

In principle, the dielectric properties of dry snow should be derivable from those of ice. Glen and Paren (1975) [4] give a brief review of mixture theory as applied to snow and firn. The equation that they prefer, and that is generally used both by them and by other authors, is known as Looyenga's equation:

$$(\varepsilon_c')^{1/3} - 1 = V(\varepsilon_i'^{1/3} - 1),$$

where V is the ratio of the density of snow to that of ice, and ε_c' and ε_i' are the real parts of the dielectric constant for snow and ice, respectively. The reasons for preferring this equation are, first, that it is the simplest to use because it is linear in $\varepsilon_i'^{1/3}$ and V, and, second, it is symmetrical with respect to which of the two dielectric components is considered as the matrix and which the inclusion, so that it is appropriate, in principle, for use across the entire density range of snow and firn.

A still simpler equation that is practically useful for radio frequencies is linearized in terms of the refractive index $n = \sqrt{\varepsilon'}$ instead of $(\varepsilon')^{1/3}$. Taking $\varepsilon_i' = 3.18$ at $-20°$ and $\rho_i = 0.917\,\mathrm{mg\,m^{-3}}$ leads to $\sqrt{\varepsilon_c'} = 1 + 0.854$ ([17], use $\varepsilon_c' = 3.17$; $\sqrt{\varepsilon_c'} = 1 + 0.851\,\rho_c$). The maximum difference relative to Looyenga's equation is 1.5% in n or 3% in ε, at $\rho_c = 0.4\,\mathrm{mg\,m^{-3}}$.

There is a difficulty with Looyenga's equation – an assumption of isotropy is made in its derivation. In fact, the value of the permittivity in snow should depend on the shapes and orientations of the particles, and the continuity of their bonding. Wiener (1910) [16] showed empirically that a single parameter, the 'Formzahl', can suffice to describe the geometrical characteristics of a mixture. Wiener's equation, applied to snow, is

$$\frac{\varepsilon_c - 1}{\varepsilon_c + u} = V\left(\frac{\varepsilon_i - 1}{\varepsilon_i + u}\right),$$

where the Formzahl, u, varies from 0 for isolated layers perpendicular to the electric field to ∞ for isolated layers parallel to the field. For spheres of ice randomly distributed in air, $u = 2$. Since polar firn is characterized by horizontal layering, Formzahls >2 are to be expected for vertically propagating electromagnetic waves.

For high frequencies, $\varepsilon'' \ll \varepsilon'$, Wiener's equation can be written separately for ε' and ε''. For ε' the form is unchanged; for ε'', the equation becomes

$$\varepsilon_c'' = \varepsilon_i'' \frac{1}{V}\left(\frac{\varepsilon' - 1}{\varepsilon_i' - 1}\right)^2.$$

Laboratory determinations of u have not been made on artificially-produced snow – those for dry natural snow yield $u \simeq 5$ for ε' and $u \simeq 2$ for ε'' (see Section 6.1).

3.7. Propagation of Electromagnetic Waves in Glaciers

The electrical properties of monocrystalline and polycrystalline ice having been examined, the problem of wave propagation in glaciers now will be considered. Glaciers may be regarded as unbounded dielectrics relative to the wave lengths used in radar sounding.

The most interesting case to consider is airborne sounding carried out from an aircraft flying at a height, H, above the surface of an ice cap of thickness, h.

In this case we may consider the ice sheet as a body of ice having extended horizontal planar interfaces, viz. the air/ice and ice/rock boundaries, penetrated by a vertically traveling pulse radiated from the airborne transmitter. To a good approximation, we may also consider the ice to be a lossless medium so that the voltage reflection coefficient at an interface may be taken to be

$$R = \frac{W - 1}{W + 1}, \tag{3.1}$$

where W is the ratio between wave impedances in the media on opposite sides of an interface, given by

$$W = \sqrt{\frac{\varepsilon_1'}{\varepsilon_2'}},$$

assuming wave propagation from medium 1, with dielectric constant ε_1', towards medium 2, with dielectric constant ε_2'.

Consequently, the power reflection coefficient becomes

$$R^2 = \left(\frac{W - 1}{W + 1}\right)^2 = \left(\frac{\sqrt{\varepsilon_1'/\varepsilon_2'} - 1}{\sqrt{\varepsilon_1'/\varepsilon_2'} + 1}\right)^2, \tag{3.2}$$

whereas the power transmitted through the interface becomes $1 - R^2$. Using $\varepsilon' = 3.17$ for ice and $\varepsilon' = 10$ for rock, we find a power reflection coefficient $R^2 = 0.079$ or $-11\,\text{dB}$ at the ice surface and $R^2 = 0.078$ or $-11\,\text{dB}$ at the base of the ice. This means that less than one tenth of the power impinging upon either interface is reflected back towards the receiver. Correspondingly, the transmission coefficient for the two interfaces becomes 0.922 or $-0.4\,\text{dB}$ for the air/ice interface and 0.921 or $-0.4\,\text{dB}$ for the ice/rock interface.

The radiation beam from antennas which may be installed on an aircraft is usually relatively wide, so that what happens to 'rays' incident upon the interface at an oblique angle will have to be considered. From Snell's law,

$$\frac{\sin \theta_2}{\sin \theta_1} = \frac{\sqrt{\varepsilon_1'}}{\sqrt{\varepsilon_2'}} = \frac{n_1}{n_2}, \tag{3.3}$$

where θ_1 and θ_2 are the angles of incidence of the incident and refracted waves, respectively, and n_1 and n_2 are the refractive indexes of the two media. The angle of refraction is smaller than the angle of incidence if $\varepsilon_2' > \varepsilon_1'$, i.e., the wave is refracted towards the normal in the medium having the larger permittivity. Taking $\varepsilon_1' = 1$ for air and $\varepsilon_2' = 3.17$ for ice, we find that

$$\frac{\sin \theta_2}{\sin \theta_1} = 0.562;$$

for instance, if $\theta_1 = 5°$, $\theta_2 = 2.8°$.

The expressions 3.1 and 3.2 for the reflection coefficients are still valid in the case of propagation at an angle oblique to the interface. There are two cases: an incidence wave (a) with the electric vector in the plane of incidence, and (b) with that vector normal to the plane of incidence. The two cases are often referred to as the vertical and horizontal polarization cases, respectively. In the first case, polarization in the plane of incidence, W assumes the value

$$W = \frac{\sqrt{\varepsilon_1'}}{\sqrt{\varepsilon_2'}} \cdot \frac{\sqrt{1 - \frac{\varepsilon_1'}{\varepsilon_2'} \sin \theta_1}}{\cos \theta_1}. \tag{3.4}$$

From (3.1) it is seen that total reflection will take place for $W = 0$, i.e., at an angle θ_{cr} given, from (3.4), by

$$\sin \theta_{cr} = \frac{\sqrt{\varepsilon_2'}}{\sqrt{\varepsilon_1'}} \qquad (3.5)$$

A real solution for θ_{cr} exists only for $\varepsilon_2' \leq \varepsilon_1'$, i.e., in the case of the air/ice interface, for waves reflected from the ice/rock interface and propagating back up through the glacier. θ_{cr} is known as the critical angle, which, in our case, becomes $\theta_{cr} = 34.1°$. The reflected wave propagates along the ice surface at the critical angle and does not reach the sounder receiver.

In the case of an airborne sounder, where transmitter and receiver use a common antenna, reflections are obtained only from bottom sections that are perpendicular to the wave front, and subglacial slopes larger than the critical angle may not be observed. In fact, power will hardly reach planes with slopes near the critical angle, since they are associated with angles of incidence of slightly less than 90°, and the radiation from the antenna in that direction is normally small.

Total reflection also occurs for $W = \infty$, i.e., for $\theta = 90°$, but this is a trivial case, since the wave then propagates parallel to the interface between two media.

In the case of polarization normal to the plane of incidence, i.e., horizontal polarization, W assumes the value

$$W = \frac{\sqrt{\varepsilon_1'}}{\sqrt{\varepsilon_2'}} \cdot \frac{\cos \theta_1}{\sqrt{1 - \frac{\varepsilon_1'}{\varepsilon_2'} \sin^2 \theta_1}}. \qquad (3.6)$$

Total reflection is obtained for $W = \infty$, i.e., for the same condition as above, and for $W = 0$, which is also a trivial case.

We have considered sounding in one direction from the antenna. Radiation from an antenna actually takes place in a certain solid angle; since each oblique 'ray' is refracted toward the normal at the air/ice interface, power density around the normal increases across the interface. This focusing produces a refraction gain ('focusing factor' in the Soviet literature) which we shall determine. Figure 19 shows an antenna at height H above thick ice that rests on rock. We assume that the air/ice and ice/rock interfaces are plane and parallel, and that the lower interface is a perfect reflector. The propagation of radio waves from the antenna through the dielectric to the lower interface and back may therefore be described by the mirror method. From the geometry of Figure 19 it is seen how the refraction at the air/ice interface creates a concentration of radiation from the antenna into a smaller area at the lower interface compared to that with no ice present (dotted 'ray'). Thus, the refraction gain may be determined as the ratio between circular areas on a reflecting interface at a distance $(H + h)$ without and with the ice. The ratio between these two radii is

$$\frac{R}{r} = \frac{(H/h) + 1}{(H/h) + (\tan \theta_2 / \tan \theta_1)}. \qquad (3.8)$$

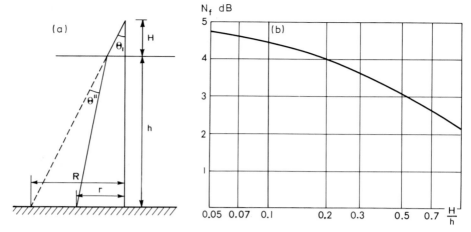

Fig. 19. a – electromagnetic energy gain due to refraction; b – quantative estimation of the refraction gain as a function of H/h (dB).

For small angles, i.e., for $\theta < 10°$ for instance, representing many practical cases, (3.8) becomes:

$$\frac{R}{r} = \frac{(H/h) + 1}{(H/h) + (\sin\theta_2/\sin\theta_1)} = \frac{(H/h) + 1}{(H/h) + (n/n_i)},$$

where $n = 1$ is the refractive index for air and n_i is that for ice.
Since $n_i > n$, $(R/r) > 1$ and the refraction gain becomes

$$N_f = 20 \log\left(\frac{R}{r}\right).$$

Figure 19 shows this gain for ice as a function of H/h and it is seen that an appreciable gain is obtained with an asymptotic value of 5 dB for $H/h \to 0$. This value is never obtained, however, since for the procedure employed it is assumed that the ice surface is in the far field of the antenna where the radiation constitutes a plane wave. Typical sounding values are $H = 300$ m and $h = 2000$ m for which the gain becomes 4.2 dB.

In the above, we have assumed that the ice sheet may be regarded as a homogeneous dielectric medium placed on top of a planar rock surface. For a general understanding of sounding conditions this is a good approximation. However, as we have seen, polar ice sheets are formed by the accumulation of annual layers of snow which are gradually compressed into solid ice. Therefore, the upper part of an ice sheet exhibits a gradual increase in density, and consequently a gradual increase in permittivity, so that the refraction properties differ from those derived above.

The general case in shown in Figure 20, with refraction caused by the sequence of horizontal layers of snow of thickness z_n and refractive index of n_n. It is assumed that the refractive index increases with depth, $n_{n-1} < n_n < n_{n+1}$, so that a 'ray bending' towards the vertical takes place.

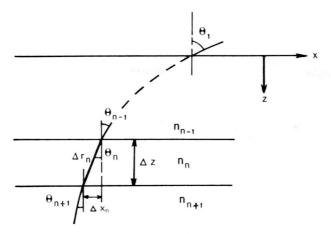

Fig. 20. Electromagnetic energy gain due to refraction in the case of propagation in layers with changing density.

Snell's law for $N - 1$ layers is:

$$\frac{\sin \theta_N}{\sin \theta_{N-1}} \frac{\sin \theta_{N-1}}{\sin \theta_{N-1}} \cdots \frac{\sin \theta_3}{\sin \theta_2} \frac{\sin \theta_2}{\sin \theta_1} = \frac{n_{N-1}}{n_N} \frac{n_{N-2}}{n_{N-1}} \cdots \frac{n_2}{n_3} \frac{n_1}{n_2}, \qquad (3.9)$$

which yields

$$\frac{\sin \theta_N}{\sin \theta_1} = \frac{n_1}{n_N}. \qquad (3.10)$$

For the sounding case, where $n_1 = 1$ (air) it is seen that the angle of refraction at the nth layer is dependent only upon the angle of incidence at the surface θ_1 and the refractive index of the layer, i.e. it is independent of all the layers above the layer considered. However, the actual path through the firn as expressed by the quantities Δx_n and Δr_n (Figure 20), is dependent upon the refractive indices and thicknesses of these layers.

From the geometry of Figure 20, one finds that

$$\Delta x_n = \Delta z_n \tan \theta_n$$

$$= \Delta z_n \frac{\sin \theta_1}{n_n} \frac{1}{\sqrt{1 - (\sin^2 \theta_1/n_n^2)}}. \qquad (3.11)$$

The total displacement from the vertical after propagation through the firn is obtained by summation over N layers.

For small angles, (3.11) may be approximated by

$$\Delta x_n \approx \frac{\Delta z_n}{n_n} \theta_1, \qquad (3.12)$$

which, for angles $\theta_1 < 10°$, underestimates Δx_n by less than 1.5%.

Similarly,

$$\Delta r_n = \frac{\Delta z_n}{\cos \theta_n} = \frac{\Delta z_n}{\sqrt{1 - \frac{\sin^2 \theta_1}{n_n^2}}} \approx \Delta z_n \left(1 + \frac{\theta_1^2}{2n_n^2}\right), \quad (3.13)$$

where the approximation overestimates Δr_n by less than 1%.

It is seen from Figure 20 that a gradual increase of refractive index results in a wider cone of radiation compared with that obtained with a homogeneous layer of ice. Consequently, a smaller refraction gain results, expressed by

$$\frac{R}{r} = \frac{(H/h) + 1}{(H/h) + (1 - (h'/h)(\tan \theta_n / \tan \theta_1) + (r'/h \tan \theta_1)}$$

$$= \frac{(H/h) + 1}{(H/h) + (1 - (h'/h)(n_1/n_n) + (r'/h \theta_1)}, \quad (3.14)$$

where h' is the thickness of the transition layer and $r' = \sum_1^N \Delta x_n$, the summation of the horizontal displacements after refraction in individual layers.

The time lapse for one passage through a dielectric medium is given by

$$t_h = \frac{r}{c_i} = \frac{r_n}{c_0}, \quad (3.15)$$

where r is given by (3.13) summed over N layers. In the radar sounding case at normal incidence, the total propagation time for a returned signal becomes

$$t_n = \frac{2}{C_0} \sum_1^N n_n \Delta z_n \quad (3.16)$$

The above formulas, which take into account a gradual increase in refractive index with depth in firn, are of interest when an exact ray tracing is desired and when the density profile is known.

For the derivation of ice thickness from radar sounding data, (3.15) is usually used, taking a velocity of $169 \, \text{m s}^{-1}$ in solid ice. Since the velocities in firn are greater, this results in an underestimation of ice thickness which may be corrected by using (3.16) for a known density profile. Several profiles have been established showing, in general, an exponential increase in density within firn layers of thickness between 30 m and 100 m with density varying between $350 \, \text{kg m}^{-3}$ and $920 \, \text{kg m}^{-3}$. Examples are:

$$\rho = 358 \exp(9.26 \times 10^{-3} z) \quad \text{with} \quad 0 < z < 100 \, \text{m},$$

and

$$\rho = 400 \exp(2.78 \times 10^{-2} z) \quad \text{with} \quad 0 < z < 30 \, \text{m},$$

the diference being dependent upon the temperature and the accumulation rate at the places in question. Using the interpretation principles mentioned above in these two examples, which are likely to represent extreme cases, the ice thickness is underestimated by about 18 m and 9 m, respectively. As to the ray tracing, the influence of firn is small. Using the first-mentioned profile it is found, for an angle of incidence $\theta_1 = 10°$, that $\Sigma \Delta x$ is about 12 m from the vertical, compared with 10 m if the ice were solid ($n = 1.78$), after having propagated through 100 m of ice, and 17.5 m if there were no ice at all.

At a depth of 2000 m, for instance, this means displacement from the vertical of 198 m, 206 m, and 352 m, respectively.

Similarly, we find that the differences in path length are small, being less than 1 m after propagation through 100 m of firn. An important conclusion of these observation is that the refraction does not introduce problems such as phase incoherence, the difference in arrival time being less than 10^{-12} s in typical cases.

An additional conclusion is that, for many practical purposes, the 'exponential' firn may be represented by a single homogeneous layer with a thickness equal to that of the firn and with a permittivity of 1.5. As to the refraction gain, it is found that the exponential firn gives rise to a reduction of the focusing factor of about 0.5 dB as compared with that calculated for solid ice.

References: Chapter 3

[1] Bullemer, B., Engelhardt, H., Riehl, N. 'Protonic conduction of ice. I. High temperature region'. In: N. Riehl et al., eds: *Physics of Ice*. New York 1969 Plenum Press, pp. 416–429.
[2] Chan, R.K., Davidson, D.W., Whalley, E. 'Effect of pressure on the dielectric properties of ice I'. *J. Chem. Phys.*, 1965, **43**, 2376–2383.
[3] Fletcher, N.H. *The chemical physics of ice*. Cambridge Univ. Press, 1970, 271 pp.
[4] Glen, J.W., Paren, J.G. 'The electrical properties of snow and ice'. *J. Glaciol.*, 1975, **15** (73), 15–38.
[5] Gough, S.R. 'A low temperature dielectric cell and the permittivity of hexagonal ice to 2 K'. *Can. J. Chemistry*, 1972, **50** (18), 3046–3051.
[6] Gross, G.W., Hayslip, I.C., Hoy, R.N. 'Electrical conductivity and relaxation in ice crystals with known impurity content'. *J. Glaciol.*, 1978, **21** (85), 143–160.
[7] Humbel, F., Jona, F., Scherrer, P. 'Anisotropie der Dielectrizitätskonstante des Eises'. *Helv. Phys. Acta*, 1953, **26**, 17–32.
[8] Johari, G.P., Charette, P.A. 'The permittivity and attenuation in polycrystalline and single-crystal ice Ih at 35 and 60 MHz'. *J. Glaciol.*, 1975, **14** (71), 293–303.
[9] Johari, G.P., Jones, S.J. 'The orientation polarization in hexagonal ice parallel and perpendicular to the c-axis'. *J. Glaciol.*, 1978, **21** (85) 259–276.
[10] Kamb, B. 'Ice polymorphism and the structure of liquid water'. In: A. Rich and N. Davidson, eds. *Structural Chemistry and Molecular Biology*. Freeman, San Francisco, 1968.
[11] Peterson, S.W., Levy, H.A. 'A single-crystal neutron diffraction study of heavy ice. *Acta Crystallogr.*, 1957, **10**, 70–76.
[12] Ramo, S. et al. *Fields and waves in communication electronics*. Chapter 6. New York, 1965.
[13] Ruepp, R. 'Electrical properties of ice Ih in single crystals'. In: E. Whalley et al., eds. *Physics and Chemistry of Ice*. Ottawa, 1973, Roy. Soc. Canada, pp. 179–186.
[14] Taubenberger, R.M., Hubmann, M., Gränicher, H. 'Effect of hydrostatic pressure on the dielectric properties of ice Ih in single crystals'. In: *Physics and Chemistry of Ice*. Ottawa, 1973, pp. 194–198.
[15] von Hippel, A., Knoll, D.B., Westphal, W.B. 'Transfer of protons through 'pure' ice Ih single crystals. I. Polarization spectra of ice Ih'. *J. Chem. Phys.*, 1971, **54**, 134–144.
[16] Wiener, O. 'Zur Theories der Refraktionskonstanten. Berichts üder die Verhandlungen der Königlich Sächsischen Gesellschaft der Wissenschaften zu Leipzig'. *Mathematisch-physikalische Klasse*. 1910, Bd. 62, Hf. 5, pp. 256–268.
[17] Robin, G. de Q, Evans, S., Bailey, J.T., 'Interpretation of radio echo sounding in polar ice sheets'. *Phil. Trans. Roy. Soc. London, Series A*, 1969, **265**, 437–505.

CHAPTER 4

EQUIPMENT FOR RADAR SOUNDING OF GLACIERS

4.1. Analysis

In this section we shall study the capability of a sounder system mounted in an aircraft flying at a certain height above the ice surface. We shall assume an idealized situation comprising a flat, homogeneous ice sheet resting on a horizontal rock bed; the interfaces are smooth planes so that no scattering takes place. We may therefore consider the measurement situation by the mirror principle (Section 3.7).

In this idealized case, neglecting the polarization losses and scattering, the received power may be expressed by

$$P_r = \frac{P_t}{4\pi[2(H+h)]^2} G_t A_r q \frac{1}{L}, \tag{4.1}$$

where P_r and P_t are the received and transmitted power, H is the height of the antenna above the ice surface, h is the ice thickness, G_t is the transmitting antenna gain, A_r is the effective area of the receiving antenna, q is the refraction gain, and L the losses involved:

$$L = L_A(L'_T)^2 L''_R, \tag{4.2}$$

where L_A is the dielectric loss by two-way propagation through the ice, L'_T is the transmission loss through the surface, and L''_R is the reflection loss at the base of the ice sheet. Note that

$$N_f = 10 \log q; \qquad N_A = 10 \log L_A; \qquad N_R = 10 \log L_R.$$

The losses may be described by:

$$L_A = \exp(2\alpha h); \qquad L'_T = \frac{1}{(1-|R'|^2)}; \qquad L''_R = \frac{1}{|R''|^2}, \tag{4.3}$$

where α is the dielectric loss in nepers per meter and $|R'|^2$ and $|R''|^2$ are the power reflection coefficients at the surface and the base, respectively. Note also that the reflection loss at the surface is

$$L'_R = \frac{1}{|R'|^2}. \tag{4.4}$$

In the airborne case, the transmitting and the receiving antennas are combined into one, and we may use a relation well known from antenna theory:

$$G = \frac{4\pi A}{\lambda^2}, \tag{4.5}$$

where λ is the wavelength of the radio wave, to study two cases: (a) the gain G is constant and independent of wavelength, and (b) the effective area A is constant and independent of wavelength.

We obtain, in case (a):

$$P_r = \frac{P_t G^2 q \lambda^2}{(4\pi)^2 [2(H+h)]^2 L} \tag{4.6}$$

and, in case (b):

$$P_r = \frac{P_t A^2 q}{\lambda^2 [2(H+h)]^2 L}. \tag{4.7}$$

In the airborne case, where the antenna is suspended under the wing or the fuselage of the aircraft, case (b) appears more realistic for choosing the frequency for optimum performance. Note also that, in (4.6) and (4.7), L_A increases with frequency.

In a pulsed sounder system we relate the receive bandwidth, B, to the pulse length of the transmitter, τ_t, by

$$\tau_t = 1/B, \tag{4.8}$$

so that the receiver noise level becomes

$$N = kTB = kFT_0 \frac{1}{\tau_t},$$

where k is Boltzmann's constant ($k = 1.38 \times 10^{-23}$ J K^{-1}), F is the noise figure of the receiver, and T_0 is the reference temperature, 290 K. Therefore, the signal-to-noise ratio at the receiver output becomes

$$\frac{S}{N} = \frac{W_t A^2 q}{\lambda^2 [2(H+h)]^2 LkT}, \tag{4.10}$$

where $W_t = P_t \tau_t$ is the energy of the transmitter, and $T = T_0 F$. In addition to L, F is slightly frequency-sensitive, increasing slowly with frequency. Equation (4.10) is the basic radar equation for reception of a single pulse with an ideal antenna system. By pulse integration an improvement in the signal-to-noise ratio can be obtained – by a factor of tf_{pr}, where t is the integration time and f_{pr} the pulse repetition frequency. Also, the antenna picks up thermal noise from the surroundings (by the main beam and its side lobes), so that the noise temperature of the antenna becomes T_a, normally different from the reference temperature T_0. Finally, the antenna cable has a loss, L_K, so that T becomes

$$T = T_0 \left(\frac{t_\alpha - 1}{L_K} + F \right), \tag{4.11}$$

where $t_\alpha = T_a/T_0$, the relative antenna noise temperature and the factor L_K^2 is included in the overall loss term L (4.10). Consequently, (4.10) becomes

$$\frac{S}{N} = \frac{W_t A^2 q g}{\lambda^2 [2(H+h)]^2 LkT}, \tag{4.12}$$

where g is the integration gain,

$$L = L_A(L_T')^2 L_R'' L_K^2, \tag{4.13}$$

and T is given by (4.11).

In (4.12), S/N stands for the signal-to-noise ratio necessary to obtain a certain probability of detection of the pulses, 50% for instance, and a certain probability of false alarm (Davis et al., 1973) [18].

Equation (4.13) may be considered as a product involving two unknowns, L_A and L_R'', the two-way dielectric absorption through the ice and reflection losses. The other two parameters, L_K and L_T are known parameters in general. Since L_A contains the distance h, the product, $L_R'' L_A$, describes the capability of a system to detect a certain surface at depth h with a reflection loss of L_R''. This product may be expressed from (4.12) as

$$L_A L_R'' = \frac{W_t A^2 qg}{(S/N)\lambda^2 4(H + h)^2 (L_T')^2 L_K^2 kT}, \tag{4.14}$$

where S/N is the minimum signal-to-noise ratio needed to obtain a certain probability of detection, or

$$L_A L_R'' = \frac{C_2 \tau_t}{(S/N)(H + h)^2}, \tag{4.15}$$

where C_2 comprises constant parameters for a certain system and a certain situation:

$$C_2 = \frac{P_t A^2 qg}{4\lambda^2 (L_T')^2 L_K^2 kT}. \tag{4.16}$$

For a particular 60 MHz system we have the following parameters: $P_t = 10\,\text{kW}$, $A = 16\,\text{m}^2$ ($G = 9\,\text{dB}$), $q = 4\,\text{dB}$, $g = 19\,\text{dB}$, $\lambda = 5\,\text{m}$, $L_T' = 0.4\,\text{dB}$, $L_K = 1\,\text{dB}$, $t_\alpha = 2.24$, $F = 2.5\,\text{dB}$ and, therefore, $T = 802\,\text{K}$. Thereby we find $C_2 = 264\,\text{dB}$. If we assume a 70 dB reflection loss at a distance of 2000 m, Figure 21 shows that we may

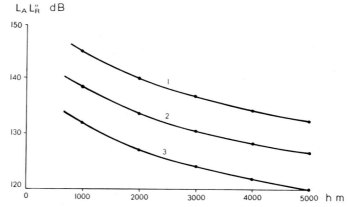

Fig. 21. Dependence of allowable loss on ice thickness. 1: $\tau = 1\,\mu s$, receiver band width $\Delta f = 1\,\text{MHz}$; 2: $\tau = 250\,\text{ns}$, $\Delta f = 4\,\text{MHz}$; 3: $\tau = 60\,\text{ns}$, $\Delta f = 14\,\text{MHz}$.

overcome a two-way loss of 57 dB if we operate the system with a pulse length of $\tau_t = 60$ ns and a receiver bandwidth of 14 MHz, whereas using a pulse length of 250 ns would allow detection of a similar reflecting surface at 4800 m if the dielectric loss here was also 57 dB. This example shows that a diagram like that of Figure 21 is a good way of describing the operational performance of a sounding system. It has the slight disadvantage, however, that some of the parameters in (4.15) and (4.16) such as A, q and L'_T only are estimates, although the estimates are well within 2 to 3 dB in total.

Occasionally, the quality of a sounding system is characterized by a number called the 'system sensitivity', which may be determined in the laboratory by measuring the loss that needs to be inserted between the transmitter output and the receiver input to give a barely detectable signal, i.e., $P_t/(P_r)_{min}$, where $(P_r)_{min}$ is the minimum permittable signal. The system sensitivity may be obtained from (4.15) and (4.16) by excluding all parameters pertaining to the antenna and the propagation of the signal, i.e., A, q, L_K, L, H, and h and setting $t_\alpha = 1$. If it is assumed that integration of received pulses is made in the same way as during field soundings, g is unchanged and with the same parameters as above we find $P_t/(P_r)_{min} = 190$ dB, 184 dB and 178 dB for the three pulse lengths 1 μs (1 MHz), 250 ns (4 MHz) and 60 ns (14 MHz).

Some of the parameters introduced above require further elaboration. The antenna noise temperature, T_a, is a measure of the thermal noise picked up by the antenna main beam and side and back lobes. In actual practice this noise is determined by the noise temperature of the ice and the cosmic noise reflected from the ice surface, weighted by the antenna gain of the main beam and the other lobes. At the low frequencies normally used for radar sounding, the cosmic noise temperature is rather high (increasing with an exponent of 2.6 with decreasing frequency). At 60 MHz the cosmic noise temperature is about 3000 K so that the cosmic contribution to T_a via the main lobe and side lobes becomes about 380 K, with a 30 K contribution by the back lobe. The ice surface has a noise temperature of 250 K (-23°C), so the total T_a becomes 660 K and $t_\alpha = 2.3$. This noise is attenuated in the antenna feed cable, which contributes noise according to $T_0(1 - 1/L_K)$, so that the noise referred to the input of the receiver can be expressed by (4.11).

The integration gain, g, is derived for the case where the received pulses are recorded on film, whereby an appreciable integration takes place. The integration is dependent upon the film speed and the spot diameter on the oscilloscope screen, the associated photographic reduction and the pulse repetition frequency. For instance, with a reduction of 1:3.33, a spot diameter of 0.2 mm and a film speed of 100 mm min^{-1}, the integration time becomes 36 ms. For a representative pulse-repetition frequency of 16 KHz 560 pulses are integrated in this period of 36 ms and, assuming a 70% efficiency, we find an integration gain of 19.5 dB (Christensen, 1970) [15]. In these considerations it is assumed that the detection is carried out by a matched filter. This cannot be realized, so some loss (about 4 dB) should be included. However, we have not taken into account the persistence of the oscilloscope screen, which is estimated to compensate approximately for the filter loss.

In case the system sensitivity is measured by observing the oscilloscope screen we will obtain another value for g. Based on experiments (Christensen, 1970) [15] we find, for a pulse-repetition frequency of 16 KHz and a 50% probability of detection, a value of only 6 dB.

4.2. Estimation of Sounding Accuracy

In the previous section we referred to the accuracy of sounding when discussing the value of refractive index and radio-wave velocity in solid polar ice. In the other sections of the book, we referred to various comparisons between seismic and radar sounding data and, at Camp Century (Greenland), between the ice thickness measured directly in the borehole and the depth determined by radar, the latter being corrected for velocities in the firn. The excellent agreement between the data in the last case is accidental, however, since numerous uncertainties are encountered in radar sounding, as we shall see presently (Gudmandsen, 1977) [26]. Uncertainties in radar sounding appear due to uncertainty in the magnitude of the refractive indices of solid ice, uncertainty in the profile of refractive index in the firn, signal contamination by thermal noise in the equipment employed, insufficient clarity of the signal on the oscilloscope, and imperfect signal definition on the recording medium. Large errors in thickness measurements can result if pulses reflected from ice inhomogeneities are mistaken for bottom echoes. The radio wave velocity is determined by the refractive index of the ice, $n = \sqrt{\varepsilon}$, where ε is the dielectric constant of ice. Laboratory measurements on solid glacier ice have yielded $n = 1.78 \pm 0.02$, so that the radio wave velocity, $c/\sqrt{n}$, becomes 168 ± 2 m sec^{-1} (Robin et al., 1969) [32]. If we assume a value of n at the uncertainty limit, we find that the thickness may be in error by about 22 m when the standard value of 168 m sec^{-1} is used as a reference and we consider a thickness of 2000 m, for instance. This means that an uncertainty in the laboratory measurement of the refractive index or a deviation in ice density of 1% will give an uncertainty of 22 m in a thickness of 2000 m, the uncertainty being proportional to the ice thickness.

In the previous chapter we discussed the influence of the gradual increase in ice density in the top hundred meters of an ice sheet. Referring to two measured density–depth profiles, it was found that a correction to the ice thickness of 18 m and 9 m should be made in the two cases, respectively. If we use a standard correction of 13 m, say, we shall have an uncertainty of about 5 m due to variations in the firn-density profile from region to region. Knowledge of these variations will reduce this uncertainty, naturally.

Normally, the radar data are recorded on film in the form of intensity-modulated echoes on an oscilloscope. In order to ensure that the smallest possible echo is recorded, the intensity modulation is adjusted so that the peak amplitude of thermal noise is just recorded (giving a faint shading on the film). This means that any echo pulse is recorded as soon as it has an instantaneous amplitude exceeding the noise level. This has a number of implications. First, the time of first appearance of the echo pulse on the film depends upon the amplitude of the pulse and the bandwidth of the receiver, which determines the pulse rise time. Second, the start of the pulse is contaminated by the thermal noise, which is added to the echo. These effects introduce an uncertainty into thickness determinations determined from intensity-modulated films (the standard method at present).

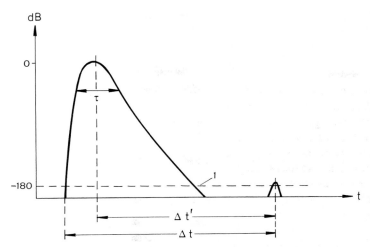

Fig. 22. Diagram to explain estimation of layer-thickness measurement accuracy. $\Delta t'$ – time interval between two pulse maxima; Δt – time difference measured on the film; 1 – radar sensitivity level.

The amplitude dependence is described schematically in Figure 22, which shows a large reference pulse (e.g., the surface echo) and a smaller echo pulse (e.g., a bottom echo) received after a certain delay. The time interval between the two pulses (corresponding, e.g., to the ice thickness) is accurately determined from the pulse peaks as shown. However, by using the intensity-modulated data a larger time interval is determined, one that depends upon the magnitudes of the amplitudes. It is seen that the ice thickness tends to be overestimated, and also that the size of the overestimation depends upon the relative amplitudes of the two pulses. Measurement in the laboratory on a specific system (pulse length $\tau_t = 250$ ns, bandwidth $B = 4$ MHz) gave the following results (Christensen, 1970) [15]:

Amplitude ratio	$\Delta t - \Delta t'$	Δh
150 dB / 60 dB	150 ns	25 m
150 dB / 140 dB	35 ns	6 m

'Amplitude ratio' means the setting of an attenuator needed to reduce the amplitude of the transmitted signal to that of the received signal (thus 150 dB corresponds to a small signal), Δt is the time difference measured on an intensity-modulated film, $\Delta t'$ is determined from an amplitude display as shown in Figure 22, and Δh is the corresponding thickness difference. The bandwidth dependence was also studied with the same equipment; the time difference as measured with bandwidths of 1 MHz ($\tau_t = 1\ \mu$s) and 14 MHz ($\tau_t = 60$ ns) was 150 ns, corresponding to 25 m in ice thickness.

These amplitude and bandwidth sensitivities give a systematic error that may, in principle, be corrected for by means of laboratory measurements (Christensen,

1970) [15]. However, since the rise time of an echo pulse is dependent upon the roughness of the reflecting surface, such correction will not be feasible in practice when considering an ice sheet surface, internal layering, or the ice/rock interface. Therefore, we must be satisfied with knowing that ice thickness may be overestimated by 10 m or 20 m when using intensity-modulation.

From Figure 22 it may be realized that thermal noise added to the pulses shown will introduce an uncertainty in the time at which the leading edge of the combined signal-plus-noise pulse crosses the recording threshold. Since this defines the start of the signal on the intensity-modulated film recording, it also introduces an uncertainty in the distance measurement. This effect is dependent upon the signal-to-noise ratio and the pulse length:

$$\sigma = 0.28 \frac{\tau_t}{\sqrt{\frac{S}{N}}}, \qquad (4.17)$$

where σ is the standard deviation of the time measurement. The effect is only of importance for small pulses; taking as an example $S/N = 1$ we find:

τ_t	σ	Δh
60 ns	16.8 ns	1.5 m
250 ns	70.0 ns	6.0 m
1 μs	280.0 ns	24.0 m

Here, Δh is the corresponding variation in depth measurement from the film recording.

Since this effect is a stochastic process by nature, the time definition of small echoes will be relatively poor when using a long pulse. This fact is also observed in practice – echoes from internal layers become rather 'woolly' when a 1 μs pulse is used.

Finally, we shall consider the uncertainty that is related to the accuracy of measurement of the position of the signal light spot on the film. If we assume a spot diameter on the oscilloscope screen of 0.2 mm, we may have an accuracy of about 0.1 mm; with a photographic reduction of 1 : 3.33 we find an accuracy of about 6 m when the depth scale is 5000 m on a 24 mm film record (Robin et al., 1969) [32].

In summary, we find that a radar-sounding survey will have an inherent uncertainty of about 15 m in depth measurement. To obtain this figure we have assumed that density variations from place to place may be considered random, with a standard deviation of 0.5%, or 11 m in depth, that variations in the density–depth profile are spatially random with a standard deviation in depth of 5 m, and that the standard deviations due to noise and spot determination are 6 m each. In addition, there is an overestimation of depth due to the dependence of the time-delay measurement on amplitude and bandwidth, which may amount to as much as 13 m.

4.3. Equipment to Measure the thickness of Cold Glaciers

All glaciers are classified according to their temperature regime and accumulation/ablation conditions into two groups: cold glaciers and temperate, or moderately cold, glaciers; the former having mean ice temperatures well below 0°C, whereas temperatures in the latter are close to 0°C.

The Antarctic and Greenland ice sheets are cold glaciers. Mountain glaciers are either cold or moderately cold, depending on their elevation above sea level. Glaciers on the archipelagos and islands in the Soviet Arctic are mostly temperate glaciers.

The differences in temperatures between these two glacier types and the periodic intense melting of temperate glaciers cause large differences in their electric properties, inner structure, and marginal conditions. It should be noted here that cold glaciers are also subject to melting at the margins. Wet snow and melt water on the ice surface significantly affect radio-wave propagation. Figure 23 shows some experimental data on the influence of surface melting on echo level. The large increase in the echo attenuation caused by the melting at the glacier surface (Figure 23) cannot be attributed solely to the higher reflection coefficient of the water surface. Thus, a layer of water ($\varepsilon' = 81$) a quarter wavelength thick on snow ($\varepsilon'_c = 2$) has a reflection coefficient $R = 0.965$, resulting in the loss of about 23 dB in the surface echo. The total echo attenuation in the glacier (N_Σ) is found to be a random value (this is confirmed by echo fluctuations); nevertheless the field data on variations in total attenuation proved to be extremely helpful in designing radar equipment for measurements on specific glaciers.

Separate terms in (1.6) can best be defined on the basis of particular special experiments.

The first experimental measurements by Soviet scientists of antarctic ice thickness were carried out in 1964. They were made in the vicinity of Mirny Observatory, with the radar 'GUYS-1M4' fitted on a surface vehicle. The ice thickness was measured only at one point on the traverse inland, 32 km from Mirny; it turned out

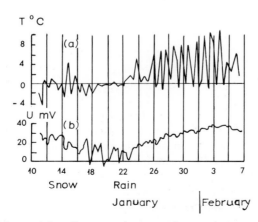

Fig. 23. Surface melting and its effect on echo strength a – air temperature during period of observation; b – amplitude of the basal reflection.

to be 850 m. The total echo attenuation for this ice thickness was estimated to be 150 dB. The first series of measurements was carried out with the radar antenna placed on the snow surface of the glacier.

In February and March of 1965 radar soundings in the Mirny area were repeated using the 'GUYS-1M4' system, which has the following specifications:

(1) Operating frequency $f_p = 213\,\text{MHz}$
(2) Pulse power $P_t = 50\,\text{kW}$
(3) Pulse length $\tau_t = 2.5\,\mu\text{s}$
(4) Pulse repetition frequency $f_{pr} = 100\,\text{Hz}$
(5) Receiver sensitivity $P_r = 5 \times 10^{-14}\,\text{W}\ (S/N = 1)$
(6) Receiver bandwidth $\Delta f = 600\,\text{kHz}$.

Radio signals are transmitted and received by a single antenna array consisting of four active element–reflector groups. The array was fixed on a cantilever beam above the sledge-mounted laboratory, toward the rear end of the sledge. The operating height of the antenna above the ice surface was 4.6 m. Echoes were received at 32 points along the Mirny–Pionerskaya traverse. To check the reproducibility of the data the measurements at two points were repeated after 30 days.

The radar sounding results were used to plot a cross-section of the ice, which was compared with one obtained by seismic shooting. The sections were found to be in a good agreement.

Experimental airborne radar sounding was carried out in February 1966. An IL-14 aircraft was equipped with the radar 'GUYS-1M4', installed in the cabin, with the antennas fixed under the fuselage. The echo signals were recorded on a photographic film from the oscilloscope screen every kilometer of the flight track by a photographic camera (of type ARFA).

The 'GUYS-1M4' radar system was used both from aircraft and surface vehicles to measure ice thicknesses ranging from 400 to 2000 m until 1976, when it was replaced by modified equipment.

To measure smaller ice thicknesses in Antarctica, an airborne radio altimeter RV-10 was employed. Its characteristics are:

(1) Operating frequency $f_p = 440\,\text{MHz}$
(2) Pulse power $P_t = 7\,\text{W}$
(3) Pulse length $\tau_t = 0.3\,\mu\text{s}$
(4) Pulse repetition frequency $f_{pr} = 10\,\text{kHz},\ 100\,\text{kHz}$
(5) Receiver sensitivity $P_r = 7 \times 10^{-13}\,\text{W}\ (S/N = 1)$
(6) Receiver bandwidth $\Delta f = 3\,\text{MHz}$.

It has a circular-sweep oscilloscope with these sweep durations: rapid sweep 10 μs, slow sweep 100 μs. It has half-wave dipole antennas with reflectors. Photographic recording was made by means of an ARFA type camera.

The RV-10 radio altimeter was used to sound glaciers with thicknesses ranging from 40 to 400 m. If the glacier temperature is $-20°$C or lower, the equipment is capable of measuring ice thicknesses of about 1000 m.

In 1967 a new radar, the RLS-60-67, was designed in the AARI, and became the main radar system for antarctic ice thickness measurements. It was in use until 1975. Its main specifications are:

(1) Operating frequency $f_p = 60\,\text{MHz}$
(2) Pulse power $P_t = 20\,\text{kW}$
(3) Pulse length $\tau_t = 0.5$ or $1.0\,\mu\text{s}$
(4) Pulse repetition frequency $f_{pr} = 1\,\text{kHz}$
(5) Receiver sensitivity $P_r = 10^{-13}\,\text{W}$ ($S/N = 1$)
(6) Receiver bandwidth $\Delta f = 1$ or $2\,\text{MHz}$.

The radar has a square-loop antenna, and two oscilloscopes, one with amplitude modulation and the other with intensity modulation. The recording from the amplitude indicator is synchronous with the photographic frame-advance rate. On the intensity indicator the recording film moves steadily across the face normal to the sweep line, so that a continuous glacier profile is recorded on the film. This radar was used to measure ice thicknesses ranging from 80 to 3300 m either from an aircraft or a surface vehicle.

The RLS-60-74 radar, developed by the AARI together with the MPI (Mari Polytechnic Institute, named after Gorky), is used at present by the SAE to measure large and medium-sized glaciers. Its specifications are:

(1) Operating frequency $f_p = 60\,\text{MHz}$
(2) Pulse power $P_t = 1$ to $60\,\text{kW}$
(3) Pulse length $\tau_t = 0.3$ to $1.0\,\mu\text{s}$
(4) Pulse repetition frequency $f_{pr} = 1$ to $3\,\text{kHz}$
(5) Receiver sensitivity $P_r = 10^{-13}\,\text{W}$
(6) Receiver bandwidth $\Delta f = 1$ or $3\,\text{MHz}$.

The radar's antenna is a half-wave dipole, log-periodic with a square frame and an active reflector. An S1-65 oscilloscope with a choice of amplitude or intensity modulation is used as an indicator for visual control. When working in the intensity-modulation mode, the oscilloscope screen is photographed on 24 mm film, which is drawn in discrete one-millimeter steps normal to the sweep line after each command impulse; the command impulses are produced at equal intervals along the vehicle track. The RLS-60-74 radar is used to measure glacier thickness from either an aircraft or a surface vehicle. It is also used to measure the velocity of surface ice movement by means of bedrock echo-amplitude variations.

To measure medium-sized ice thicknesses, the 60-72 radar was developed in the AARI. Its specifications are:

(1) Operating frequency $f_p = 60\,\text{MHz}$
(2) Pulse power $P_t = 1\,\text{kW}$
(3) Pulse length $\tau_t = 0.1\,\mu\text{s}$
(4) Pulse repetition frequency $f_{pr} = 10\,\text{kHz}$
(5) Receiver sensitivity $P_r = 5 \times 10^{-13}\,\text{W}$
(6) Receiver bandwidth $\Delta f = 7\,\text{MHz}$.

Its antenna is a half-wave dipole. The signal indicators are identical to those of the RLS-60-74 radar. The RLS-60-72 radar is used to sound ice shelves and coastal glaciers. Normally it is mounted either on an aircraft or a helicopter. It was used to measure ice thickness on the Lazarev, Amery, Shackleton, and Filchner Ice Shelves.

In 1976 a new radar, RLS-60-76, for sounding small ice thicknesses was developed in the AARI. The main specifications of the RLS-60-76 are:

(1) Operating frequency $f_p = 60\,\text{MHz}$
(2) Pulse power $P_t = 100\,\text{W}$
(3) Pulse length $\tau_t = 0.03\,\mu\text{s}$
(4) Pulse repetition frequency $f_{pr} = 20\,\text{kHz}$
(5) Receiver sensitivity   not determined
(6) Receiver bandwidth $\Delta f = 35\,\text{MHz}$

The antenna is a broadband dipole which moves on the ice surface behind the vehicle. Indicators and recorders are the same as in the RLS-60-74 radar.

The RLS-60-76 gives reliable results in the thickness range of 3 to 200 m.

4.4. Equipment to measure the thickness of temperate glaciers

Temperate (or moderately cold) glaciers are characterized by a more heterogeneous inner structure than that of cold glaciers. The layers in temperate glaciers have stronger density contrasts due to repeated melting of ice and freezing of melt water and wet snow on the surface. In addition, temperate glaciers contain more morainal material and impurity layers caused by rock weathering. In the warmer ice of these glaciers radio-wave absorption is greater than in cold glaciers.

Table VI shows the absorption of plane-wave energy during propagation through a glacier 1000 m thick with ice temperatures characteristic of cold and moderately cold glaciers.

TABLE VI

The absorption (in dB) of plane-wave energy of different frequencies within glaciers at temperatures $-1°C$ and $-20°C$

Temperature (°C)	Frequency (MHz)					
	440	200	100	60	30	10
-1	102.4	97.8	91.2	87	81	70
-20	40			30		

A layer of wet snow or water on the ice surface will result in an echo attenuation as great as 20 dB (see 4.3).

Experiments have shown that inhomogeneities in the ice cause signal attenuation of 10 dB or more. The attenuation has been found to depend on both the number of inhomogeneous layers with inhomogeneities and the dimensions of the inhomogeneities that scatter the signal.

The standard radar equipment usually used for cold glacier sounding does not operate adequately on temperate glaciers, because of their structural features and regime.

Since the maximum thickness of moderately cold glaciers normally does not exceed 600 m, the operating frequency of the equipment can be decreased to 10–100 MHz to lower the specific absorption and losses due to scatter, or increased to 600–1000 MHz, frequencies that are associated with antennas having high gain coefficients and narrow beams.

Field measurements have shown that the RV-10 radar can be used for sounding of glaciers that are not more than 200 m thick. To study glaciers with greater thicknesses, equipment with a higher operating frequency (up to 100 MHz) is used, as well as the radar RLS-60-72.

To measure thicknesses of mountain glaciers the following equipment was developed in the MPI.

Radar RLS-76; its specifications are:

(1) Operating frequency f_p = 490–930 MHz
(2) Pulse power P_t = 8–25 W
(3) Pulse length τ_t = 0.3–1 μs
(4) Pulse repetition frequency Not given
(5) Receiver sensitivity P_r = 10^{-10} dB
(6) Receiver bandwidth Δf = up to 15 MHz.

This radar has a log-periodic antenna. Recording is continuous on 35 mm movie film. The radar was used for thickness sounding of glaciers in the polar Urals.

Radar RLS-77 (used on Spitsbergen glaciers, mounted on a helicopter MI-8). Its specifications are:

(1) Operating frequency f_p = 620 MHz
(2) Pulse power P_t = 40 W
(3) Pulse length τ_t = 0.3 μs
(4) Pulse repetition frequency Not indicated
(5) Receiver sensitivity P_r = 10^{-12} dB
(6) Receiver bandwidth Δf = 5 MHz

The antenna is a grid consisting of 16 (4 × 4) reflectors of a triple-square type, with a beam width of 18° (to half-power points) in E and H planes. The recording from the intensity-modulating indicator is continuous on a 35 mm movie film.

4.5. Radars to Study Internal Structure and State

'Introscopy' of glaciers, or the study of their internal structure and state, has been carried out in Antarctica since 1967 (12th SAE). Radars are used to detect layers, inhomogeneities of different origins, voids, and crevasses in glaciers. They are also used to recognize areas with increased stresses, and to estimate mean glacier temperature.

Special-purpose radars have been developed for introscopy and new methods of observation have been tested. In particular, two-frequency sounding has proved valuable for revealing near-bottom stratification, while polarization studies offer advantages for revealing stress states in the ice.

The characteristics of the radars used for the study of ice internal features are given below.

Radar RLS-100:

(1) Operating frequency $f_p = 100\,\text{MHz}$
(2) Pulse power $P_t = 7\,\text{W}$
(3) Pulse length $\tau_t = 0.3\,\mu\text{s}$
(4) Pulse repetition frequency $f_{pr} = 100\,\text{kHz}$
(5) Receiver sensitivity $P_r = 10^{-12}\,\text{W}\ (S/N = 1)$
(6) Receiver bandwidth $\Delta f = 3\,\text{MHz}$

Its antenna is a half-wave dipole; the oscilloscope has amplitude modulation and a linear sweep.

In 1967 this radar, together with the RV-10, was successfully used in coastal antarctic areas for locating and investigating internal (near-bottom) layers containing morainal material.

Measurements of total attenuation at two frequencies made by means of the RV-10 and GUYS-1M4 radars in 1966 displayed agreement with theoretical (calculation) results. (See 1.5.) This led to the conclusion that it would be possible to estimate the mean temperature of a glacier. Thus, for instance, the experimentally-obtained value of $\tan\delta = (5-7) \times 10^{-4}$ ($f_p = 213\,\text{MHz}$) suggests that between the 25th and the 100th kilometer along the Mirny-Pionerskaya traverse the mean temperature of the ice should be $-20°\text{C}$. This is in a good agreement with the temperature measurements in the hole drilled in the area.

In 1967 in the AARI, a new five-frequency radar, the RLS-5-67, was developed. It has the following specifications:

(1) Operating frequency $f_p = 70, 100, 140, 220, 440\,\text{MHz}$
(2) Pulse power $P_t = 7\,\text{W}$
(3) Pulse length $\tau_t = 0.3\,\mu\text{s}$
(4) Pulse repetition frequency $f_{pr} = 100\,\text{kHz}$
(5) Receiver sensitivity $P_r = 10^{-12}\,\text{W}\ (S/N = 1)$
(6) Receiver bandwidth $\Delta f = 3\,\text{MHz}.$

The radar has a half-wave dipole antenna and an amplitude-modulated oscilloscope with linear sweep. This radar is designed for the estimation of the total attenuation of radio signals and investigation of frequency relationships. The measurements made by means of RLS-5-67 radar confirmed that the total attenuation of radio signals in the antarctic glacial ice had a very weak dependence on frequency.

To study depolarization of radio waves in glaciers, a radar with a narrow antenna beam was developed (RLS-440-72 radar):

(1) Operating frequency $f_p = 440\,\text{MHz}$
(2) Pulse power $P_t = 500\,\text{W}$
(3) Pulse length $\tau_t = 0.1\,\mu\text{s}$
(4) Pulse repetition frequency $f_{pr} = 10\,\text{kHz}$

(5) Receiver sensitivity $\quad P_r = 3 \times 10^{-12}$ W $(S/N = 1)$
(6) Receiver bandwidth $\quad \Delta f = 10$ MHz.

It has a rotating antenna, with in-phase fed elements and a reflector, giving a narrow beam (20° to half-power points.) The signal is recorded by a camera that rotates synchronously with the antenna, producing a circular scan on the photographic record. The bedrock or layer echo is displayed on the film as a circle made of points whose brightness depends on the amplitude of the echo, which changes as the angle between the transmitting and receiving dipoles changes.

This radar was also used to study the pattern of back scatter of the bedrock echoes and for radar surveying of the ice bottom.

Safety in traveling over the ice surface requires reliable detection of crevasses that are more than 1 m wide. To detect and study ice crevasses the RV-10 radar was modified, so becoming the RLS-440-69 radar:

(1) Operating frequency $\quad f_p = 440$ MHz
(2) Pulse power $\quad P_t = 10$ W
(3) Pulse length $\quad \tau_t = 0.1$ μs
(4) Pulse repetition frequency $\quad f_{pr} = 100$ kHz
(5) Receiver sensitivity $\quad P_r = 10^{-11}$ W $(S/N = 1)$
(6) Receiver bandwidth $\quad \Delta f = 10$ MHz.

Its antenna is similar to that of the RLS-440-72 system.

In the Riga Civil Aviation Institute special equipment for crevasse detection was developed.

The following are the main characteristics of these radars.

Radar RS-1:

(1) Operating frequency $\quad f_p = 10$ GHz
(2) Pulse power $\quad P_t = 1$ kW
(3) Pulse length $\quad \tau_t = 10$ to 15 ns
(4) Receiver sensitivity $\quad P_r = 10^{-9}$ W $(S/N = 1)$.

This radar has a horn or a parabolic reflector antenna with beam width 20° and 10°, respectively.

Radar RS-2

(1) Operating frequency $\quad f_p = 250$ or 440 MHz
(2) Pulse power $\quad P_t = 150$ W
(3) Pulse length $\quad \tau_t = 10$ to 12 ns
(4) Receiver sensitivity $\quad P_r = 10^{-7}$ W $(S/N = 1)$.

It has a wide-band half-wave dipole antenna with reflectors and two directors.

The sounding pulses at 250 and 440 MHz (pulse lengths 10–12 ns) are formed by shock excitation of the antenna, whereas at 10 GHz they are formed by a magnetron generator with a pilot excitor. The recording was made by means of a gated indicator after pre-amplification and detection of echoes.

In 1980, the AARI-developed equipment for measuring annual snow layers was successfully tested in Antarctica. Echoes from density anomalies in the snow/firn layer were detected, the maximum depth being 10 m. The following are the specifications of the equipment used:

(1) Operating frequency $f_p = 10\,\text{GHz}$
(2) Pulse power $P_t = 10\,\text{W}$
(3) Pulse length $\tau_t = 1\,\text{ns}$
(4) Pulse repetition frequency $f_{pr} = 20\,\text{kHz}$
(5) Receiver sensitivity $P_r = 10^{-7}\,\text{W}\ (S/N = 1)$
(6) Receiver bandwidth $\Delta f = 2\,\text{GHz}$.

The receiving and transmitting antennas (horn type) were mounted on a Kharkovchanka-2 tractor, 1 m apart.

The echoes after amplification and gating, were fed into a displaying and recording unit similar to that of the RLS-60-74 radar.

4.6. Radars for Ice Movement Measurements

Nowadays the surface velocity of glaciers can be measured by radar sounding and laser techniques. The radar technique is based on the comparison of signals reflected from different bedrock features. The echo returned from the rough bedrock is known to consist of a sequence of pulses. If the antennas of the radar move horizontally, along the ice surface, the distances to the bedrock shape that forms the echoes change. This leads to variations of the phase relationships between pulses in a group, and thus to changes in their envelope. For small distances these changes are quasi-periodic and are called spatial fading. Since the pattern of amplitude variations was found to be mainly controlled by the glacier bed roughness, a repeated survey along the same line after a certain time interval enables the displacement of the glacier surface relative to its base to be determined, if the measuring line is along the direction of motion. It should be noted here that the radar method does not require any reference to rock outcrops, and hence can be used almost everywhere in Antarctica except on the ice shelves, where the echo is not subject to fading.

To measure surface velocity the RLS-60-74 radar was used (See 4.3). The receiving and transmitting antennas were both logperiodic. They were mounted on opposite sides of a Kharkovchanka vehicle. To get the pattern of amplitude changes, the echo, after intensity modulation and expansion to the full width of the oscilloscope screen, was photographed by a fast photographic camera, the 'RFK-5'.

The film was drawn synchronously with the antenna movement. The absolute error in the velocity in ice has never exceeded 0.1 m [sic].

4.7. Laser Technique

Laser measurement of the velocity in ice is based on the Doppler effect. By the variations in beat frequency of a coherent signal, a velocity component of the reflector movement can be determined. The reflector should be fixed rigidly to the ice surface. The measuring equipment should be placed on a massive foundation, stable relative to the glacier. In 1968 the prospects of Doppler laser systems were discussed with regard to the measurements of glacier flow velocity by Belousova et al., (1971) [2] and Bogorodsky et al., (1970) [6]. The Doppler frequency (beat frequency) is known to be determined by

$$\omega_D = \omega_0 - \frac{2v}{c_0}\omega_0,$$

where ω_0 is the laser frequency, c_0 is the speed of light, and v is the velocity of the glacier flow. Bogorodsky et al. (1974) showed that the input power of the laser receiver ($P(t)$) is given by

$$P(t) = \tfrac{1}{2}A^2(t) = A^2[1 + \cos(2\omega_D t)],$$

where A is the amplitude of the laser vibrations. The power ($P(t)$) varies from 0 to $2A^2$ with frequency $2\omega_D$. Table VII lists Doppler frequencies for a laser operating in the green region of the optical spectrum, for various glacier velocities.

TABLE VII
Doppler frequencies for various glacier velocities

Glacier velocity		Doppler frequency (Hz)	Doubled Doppler Frequency (Hz)
(m yr^{-1})	(mm s^{-1})		
1.58	5×10^{-5}	0.1	0.2
15.8	5×10^{-4}	1.0	2.0
158.0	5×10^{-3}	10.0	20.0

As seen from Table VII, it takes 50 s to measure the velocity of a slowly moving glacier (1.58 m yr^{-1}), i.e., to provide 10 cycles of output to the recording instrument. In principle, even lower velocities can be recorded, such as a few centimeters a year. This would permit measurements of glacial strain rate to be made by means of a laser.

In 1970 (15th SAE) the first extensive experiments were carried out. The Doppler laser velocity meter had the following specifications: it operated on a mixture of helium and neon (He–Ne laser), at a wavelength of 632.8 nm. A beat frequency (ω_D) of 3.2 Hz corresponds to a glacier velocity of 1 μm s^{-1}. To get 100% modulation of a signal at 3.2 Hz, a laser generating a single mode at a single frequency was used. A model 13 He–Ne gas laser is the main unit of the Doppler laser velocity meter. The angle of laser radiation divergence is 1.2×10^{-3} radians. To get a better antenna beam, a 16-power telescopic system that provided for a net divergence angle of 7.5×10^{-5} radians was used. The silvered mirror-reflector had a reflection coefficient of 96%. Its diameter was 20 cm. A FEU-36 photomultiplier tube, sensitive to the range of 300–600 nm, was used as a receiver. An interference light filter with $\lambda = 632.8$ nm and a bandwidth of 2.4 nm was placed at the output of the photomultiplier tube (to reduce background noise). The reflecting mirror was placed directly on the ice and was protected from the wind by a metallic case. Special tests have shown that working conditions were most favorable when no atmospheric disturbances occurred at the time (eddy air fluxes due to ground heating, wind, and the like). The distance between the laser and mirror sometimes reached 1.8 km.

The equipment for ice velocity measurements by laser technique has the following specifications.

Range of velocities measured: $0.5-500\,\mu\mathrm{m\,s^{-1}}$;
smallest displacement recorded: $0.1\,\mu\mathrm{m}$;
maximum distance to the studied features: 1.5 km;
laser wave length $\lambda = 632.8$ nm;
angle of beam divergence = 2'.

The laser used in these tests operates at a single frequency and generates a single mode. A triple prism reflector is mounted on the studied object. Its light spot diameter is 160 mm, and its accuracy (as manufactured) is 2 ±5″. The reflector rotates ±10° in the vertical plane, and ±180° in the horizontal plane.

4.8. Equipment of U.S.A., Denmark, and Great Britain for Radar Sounding of Cold Glaciers

The first radar soundings of ice thickness were carried out using the standard U.S. Air Force radio altimeter called the SCR-718 (Waite and Schmidt, 1962) [42], equivalent to the Soviet RV-10. This is a conventional pulse-modulated radar system operating with a 3 MHz bandwidth about a center frequency of 440 MHz. As it was designed to produce reflections off a very large target, the Earth, it has only 7 W peak pulse power. Consequently, it is light in weight and consumes little power. The outgoing pulse and reflection are displayed on a circular sweep on a cathode ray tube. A further description of the instrument is given in Sinsheimer (1947) [36]. Because it lacks the high penetration power of specially-built systems, an unmodified SCR-718 is limited in measurement capability to depths less than 500–1000 m, depending on temperature. The reading accuracy on the display is ±15 m as measured in ice, and the minimum sounding depth owing to the length of the outgoing pulse is about 50 m.

In Canadian measurements the SCR-718 was mounted in a cabin on a Nansen sled pulled by a motor toboggan. Dipole antennas were mounted with the transmitting antenna on one side of the sled and receiving antenna on the other, oriented perpendicular to the direction of travel. The antennas were backed by aluminum sheet reflectors. Power was supplied by two 12 V vehicle batteries which drove a 110 V/400 Hz dynamotor. Batteries were either continuously charged, or recharged overnight after each day's soundings. In the experiments of Weber and Andrieux (1970) [47], the vehicle had to be moving to obtain a depth measurement, because the bottom reflection could be distinguished from a multitude of other events occurring on the screen only through its recurrence. Paterson and Koerner (1974) [30] found, on the other hand, that there were few reflections observed from layers within the ice and that there was no difficulty in distinguishing the bedrock reflection while the vehicle was stationary.

Paterson and Koerner (1974) [30] also employed a continuously-recording system which permitted ice thickness to be displayed in relation to distance on a storage oscilloscope which was photographed at the end of a run. The image on the oscilloscope lasted for 10 or 15 min, in which time a distance of about 3 km could be sounded. Unfortunately, the power consumption of the system was so high that the batteries were discharged in about an hour, so it was seldom used.

Further use of the SCR-718, mounted on man-haul sledges, was made on Roslin Gletsher in East Greenland (Davis et al., 1973) [18]. Two antennas of the rectangular 'trough' type, 1.5 wavelengths wide, 1 wavelength long, and with corner angles of 45° were used. The measured forward gain of each antenna was 8 dB. Early in the 1960s, recognizing the limitations of the radar altimeter, Waite at the U.S. Army Electronics laboratory and Evans at the Scott Polar Research Institute separately developed systems designed specifically for sounding of ice (Evans, 1963) [21]. The first system developed by Waite, USAEL Mark I, operated on a center frequency of 30 MHz, with 400 W peak pulse power, a pulsewidth 0.4 s, and a pulse repetition frequency of 2 kHz. The receiver had a bandwidth of 4 MHz, a gain of 130 dB and fed into a dual-trace Tektronix type 45A oscilloscope. Pictures of the oscilloscope screen were taken with a motor-driven 35 mm camera that was mechanically adjustable to film speeds of 0.2 to 8.5 mm s^{-1}, and therefore adaptable to continuous profiling at either surface or aircraft speeds. Antennas were broad-band 30 MHz folded-dipoles mounted end-to-end on opposite sides of the vehicle or aircraft.

In 1963, Waite developed the USAEL Mark II system. The transmitter generated half-microsecond pulses at a nominal center frequency of 30 MHz with a repetition frequency of 20 kHz. Peak output power was 300 W at a nominal line voltage of 115 VAC. Variation in peak output due to supply voltage change was 0.08 dB V^{-1} over the range 100 to 130 VAC. The receiver provided linear-logarithmic response to 30 MHz pulse signals, had a bandwidth of 3 MHz, and a noise figure of less than 2 dB. A transmit/receive switch, which permitted the use of one antenna, provided isolation of over 40 dB between transmitter and receiver and insertion loss of less than 0.3 dB to the transmitted and received signals. System waveforms and returned pulse delay times were observed on a dual-trace oscilloscope. The USAEL Mark II system was used in a comparative experiment in Greenland (Rinker and Mock, 1967; Evans, 1967) [31, 20] in sounding flights over Greenland (Walker et al., 1968) [45] and in the early experiments by the University of Wisconsin-Madison in Antarctica (Figure 24), including the first echo sounding on the East Antarctic plateau, at South Pole Station in January, 1965 (Jiracek, 1967) [50].

The equipment first developed by Evans, called the SPRI Mark I, was designed to operate at 35 MHz with a peak pulse power of 80 W, a bandpass of 14 MHz, and a p.r.f. of 50 kHz. The Mark I used in Antarctica had an oscilloscope presentation similar to the SCR-718, but it was also provided with a chart recorder. The SPRI Mark I system was soon succeeded by Mark II, which is fully described by Evans and Smith (1969) [22]. The main differences from the Mark I system was that receiver and control circuits were transistorized, and a continuously moving photographic film was used to record echoes displayed on an intensity-modulated cathode ray tube.

Fig. 24. Equipment used in early Antarctic radar sounding by G.R. Jiracek of the University of Wisconsin. Top: wannigan housing radar equipment pulled by motor toboggan. Bottom: USAEL Mark II system mounted inside wannigan. (Jiracek, 1967).

The SPRI Mark II system has become widely used, because it was turned over to a commercial firm for manufacture and sale at a reasonable cost. It has remained the basic unit employed in the airborne systems that produced most of the soundings of the Antarctic ice sheet by the U.S.A., Great Britain, and other western countries up until about 1970, when it was replaced by the more modern system developed by Gudmandsen in Denmark. It is still widely used to this day.

The first field test in Antarctica of the SPRI Mark I system was in December 1963 (Walford, 1964) [43]. The equipment was mounted on a tractor with separate transmit and receive unipole antennas, which transmitted and received the most power in their equatorial planes (i.e., the planes containing the vertical and the direction of travel). Due to refraction, the sensitivity in the ice is greatly reduced beyond 35° to the vertical. There was no other means provided for reducing the sensitivity to echoes received off nadir, or for detecting the direction from which the echoes arrived.

The Mark II system was first tested in Greenland in 1964, and then installed again on a tractor for further measurements in Antarctica. For these experiments (Bailey and Evans, 1968) [10] a different disposition of the antennas was used. The transmitting antenna was mounted forward from the cab of the tractor and in line with the direction of travel. The receiving antenna was supported on a sled 18 m away from, and in line with, the transmitting antenna, thus reducing the strength of the signal traveling directly from transmitter to receiver by approximately 50 dB. No difficulties with sensitivity were found from having a long length of coaxial cable, other than the inconvenience and the problem of preventing damage at towbar connections between sleds.

The SPRI Mark II system was used on extensive oversnow traversing in Queen Maud Land by the University of Wisconsin-Madison in 1965–66 (Beitzel, 1971) [13] and 1967–68 (Clough *et al.*, 1968) [17]. The electronic units were placed in the cab of a large Model 843 Sno-Cat, separated by some 20 m from a specially constructed antenna sled at the rear of the sled train. The antennas, which were mounted parallel to each other, were rigid folded dipoles, 5 m long, each with a single reflector element 2 m above the antenna. The reflectors were found to enhance the signal by 3 to 5 dB. Continuous recording was not available for this system; instead the record of the travel time observed on the oscilloscope was manually recorded at frequent intervals. Near the end of the 1967–68 season, field tests were successfully carried out on a prototype digital profiling system.

The South African Antarctic Expeditions have also used the SPRI Mark II system for several seasons of oversnow work (Barnard, 1975; Schaefer, 1973; Van Zyl, 1973 [11, 34, 41]. The majority of soundings were made with a single folded-dipole antenna mounted 2 m above the snow surface parallel to the track direction, and a transmit/receive switch (Schaefer, 1972) [35]. Two oscilloscopes were employed, one for amplitude-modulation monitoring, and one for continuous intensity-modulated recording. In 1972, a second antenna was mounted and fed in parallel to the first, increasing the vertical gain by about 3 dB and cutting down on some of the side lobes (Van Zyl, 1973) [41].

The SPRI Mark II system was first mounted in a light aircraft for measurements over Ellesmere Island in 1966 (Evans and Robin, 1966) [23]. Measurements with similar mountings soon followed in Antarctica (Swithinbank, 1968; Van Autenboer and Decleir, 1971) [39, 40]. Use of the Mark II, or the similar Mark IV, in light aircraft (principally a twin turbo-prop DeHavilland 'Twin Otter') has continued up to the present, especially by the British Antarctic Survey (Smith, 1972) [37] and the Ross Ice Shelf Geophysical and Glaciological Survey (RIGGS) (Bentley *et al.*, 1979) [14]. The Mark IV system has a higher transmitter power than the Mark II, an improved receiver recovery characteristic, and can annotate the 35 mm film record with date, time, receiver gain, and display parameters. The antenna mountings have varied. In their early work, the British Antarctic Survey used a half-wave folded-dipole whip antenna clamped to each wing strut, one side being used for transmitting and the other for receiving (Swithinbank, 1968) [39]. In their later work (Smith, 1972) [37], the antenna system consisted of a folded wire dipole 6 m long, centered on an impedance-matching unit fixed to the belly of the aircraft. The impedance-matching unit contained a network which allowed the antenna to be fed from a 50 ohm coaxial cable, and also a terminating resistor to load the center of the dipole. Nylon cord joined the ends the dipole to the tips of two 1 m steel masts, which extended down from the wing tips. The 3 dB beam width perpendicular to

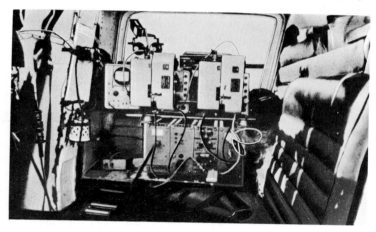

Fig. 25. Mounting of a SPRI Mark IV radar in (a) and on (b) a helicopter for soundings in Svalbard. (Drewry *et al.*, 1980).

the flight path was 20°. The antenna gain was only about $-10\,dB$, presumably because of the proximity of the skis and body of the aircraft. An essentially identical mounting was used by Van Autenboer and Decleir (1969) [40]. The antennas used in the RIGGS program were folded dipoles constructed of flexible wire, held taut under each wing by nylon cord. The antennas were subparallel to the wings, with the transmitting antenna under one wing and the receiving antenna under the other. The electrical characteristic of the antennas were not measured. Recently, soundings have again been carried out from a helicopter (Drewry et al., 1980) [19]. A SPRI Mark IV system, now operating with a center frequency of 60 MHz, was used for measurements in Svalbard. It was mounted in a Bell-206 helicopter, with one monitoring and two recording oscilloscopes (Figure 25). A simple dipole antenna was mounted parallel to the flight direction on a float. Results were recorded in standard manner on 35 mm film. A similar arrangement was used in 1978–79 in Antarctica (Orheim, 1980) [29].

The Australian National Antarctic Research Expeditions have developed their own sounder (Morgan and Budd, 1975) [27]. Operating on a center frequency of 100 MHz, it has a peak pulse power of 5 kW and an overall system pertormance of 175 dB. In has been used both for ground surveys, with motor toboggans and dog sleds, and for airborne sounding.

In 1967–68 the cooperative program between the Scott Polar Research Institute and the U.S. National Science Foundation to undertake the sounding of the Antarctic ice sheet from long-range aircraft began. The installation of the equipment has been described in detail by Evans and Smith (1969) [22].

4.9. Equipment of U.S.A., Denmark, and Great Britain for Radar Sounding of Temperate Glaciers

Temperate glaciers are more difficult to sound than cold glaciers for two reasons, (1) the higher attenuation in warm ice, and (2) – and probably more important – the scattering caused by discontinuities within the ice. The approach to the sounding of temperate glaciers has been two-fold: to increase the operating frequency, and to decrease the operating frequency. UHF (~1 GHz) sounders provide improved directionality, whereas low frequencies give wavelengths long enough to reduce scattering.

A radio frequency interferometric technique developed initially for use with lunar exploration was applied on Athabasca Glacier in 1973 (Strangway et al., 1974) [38]. In this set-up, a horizontal electric dipole was laid out on the surface and used to transmit electromagnetic energy of frequencies of 1, 2, 4, 8, 16, 32 MHz in sequence. A receiving coil mounted on a vehicle was moved away from the transmitting antenna and the field strength of each frequency was detected and recorded on magnetic tape and on a strip chart recorder. The interference pattern produced is dependent upon the glacier depth, and therefore could be used as a depth measurement.

In the instrumentation of Strangway et al. (1974) [38], two orthogonal transmitting antennas and three orthogonal receiving coils were employed. By transmitting and receiving with each of the possible combinations in sequence, six separate pieces of information were recorded at each frequency, some with maximum antenna coupling, containing the interference patterns, and some with minimal coupling – the latter are potentially useful as indicators of scattering from the subsurface or reflections from lateral, known inhomogeneities such as valley walls. Field data were interpreted by matching with theoretical curves.

The main thrust in the development of sounders for temperate glaciers, however, has been to use radar pulse systems with their characteristics appropriately modified. The first system specifically designed for temperate glaciers was that of Goodman (Goodman and Terroux, 1973; Goodman, 1975) [25, 24]. A 620 MHz frequency (which is in the UHF television band) was chosen, since components and engineering data were readily available. In tests on Athabasca Glacier and Wapta Ice Field in the Canadian Rockies, Goodman (1975) [24] found that scattering at that high a frequency from glacial inhomogeneities, such as bubble–ice layers and water-filled cavities, was not a hindrance, but rather yielded additional data concerning glacial structure. The main design objectives were short pulse lengths, high peak pulse power, and a capability of handling a large dynamic range of reflected signals. The radar was to be of solid-state construction to reduce power and weight and achieve increased reliability.

The following description is quoted from Goodman (1975) [24].

The radar set is a conventional design consisting of three components: the transmitter, the receiver, and the antenna which includes a send-receive switch.

The *transmitter* [Figure 26] is a grid-modulated cavity oscillator using a 'light-house' output stage and a

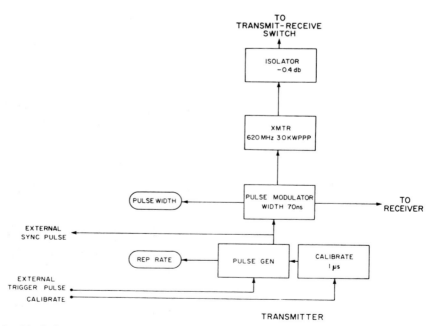

Fig. 26. Block diagram of 620-MHz radar transmitter for temperate glacier sounder (Goodman, 1975). Transmit–receive switch; isolator – 0.4 dB; transmitter – 620 MHz, 3 KWt peak pulse power; pulse modulator width 70 ns; calibration unit, 1 μs.

low-Q tank circuit to give a peak pulse power of 3 kW. An isolator prevents transmitter self-destruction in the event of loss of load. The pulse modulator has a variable pulse rate (1 pulse per second to 20 000 pulses per second) and three possible widths (70, 250, and 500 ns).

The *receiver* [Figure 27] is a superheterodyne type with an input pre-amplifier that has a noise level of 3 dB above thermal, a dynamic range of 90 dB and a gain of 10 dB. A 100 MHz logarithmic IF amplifier with 30 MHz band width and 90 dB dynamic range provide most of the receiver gain (80 dB). A conventional video detector and amplifier complete the receiver.

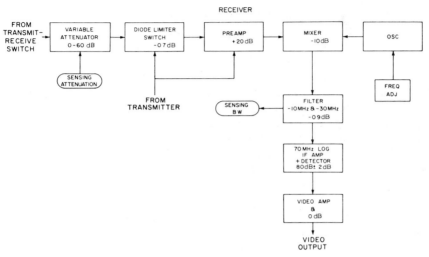

Fig. 27. Block diagram of radar receiver for temperate glacier sounder (Goodman, 1975).

Fig. 28. Antenna for temperate glacier sounder (Goodman, 1975).

The *antenna* [Figure 28] is a corner reflector which was chosen in order to achieve good spatial resolution and small side lobes. The beam width is 5.2° and the gain is 15 dB. These excellent characteristics required a rather bulky antenna, consisting of a triangle 2 m across its base and 2 m high. A transmit–receive switch is preferable to the alternative of two antennas. The receiver is protected by a set of fact-recovery diodes and the main isolation transmit–receive switch is a ferrite circulator. Recovery times in the order of 100 ns are achieved, with 60 dB isolation between transmitter and receiver.

A precision calibration circuit which generates two continuous transmit pulses with 1 μs spacing is an integral part of the radar system. Since both pulses use the same electronics, all delays are compensated and the calibration is independent of transmitter and receiver characteristics.

One of the big advances of Goodman's system was the incorporation of a computer data handling capability, further described in Chapter 5. Digital recording techniques greatly increased the recording resolution and provide the results in a form directly compatible with a computer. A UHF radar system similar to the Goodman system has been developed recently at the University of British Columbia (Narod and Clarke, 1980) [28]. Although designed specifically for airborne sounding of the polar glaciers and ice caps of northern Canada, it presumably could also be used on temperature glaciers. Its center frequency is 840 MHz, bandwidth 40 MHz, and system performance is 124 dB. The 35% increase in frequency over that of the DOEC system brings increased susceptibility to internal scattering and higher dielectric losses, but these drawbacks are offset by the high gain and small size of the antenna and the ease with which the entire system is made airborne. By moving the antenna while continuously recording, the relatively constant bottom echo can be identified among the highly variable scattering returns. When airborne, the scattering returns vary too rapidly to be visible on an oscilloscope phosphor.

The UBC echo sounder comprises a transmitter, receiver, circulator, antenna, and oscilloscope display. Adding the two-way antenna gain of 31 dB, the effective system performance is 155 dB. Both the transmitter and receiver are internally protected against abnormal power surges caused by cable failures. A 120 MHz signal from a crystal oscillator is multiplied to 720 MHz and 840 MHz to provide the local oscillator and carrier frequencies, respectively. The carrier is then amplified, gated, and isolated at the 4 kW level. The transmit–receive switch is a three-port ferrite device which allows the system to operate with a single antenna. The receiver input is protected by a solid-state limiter. The signal is converted to the 120 MHz intermediate frequency and filtered. The I.F. amplifier–detector has a logarithmic response characteristic with a useful dynamic range of 80 dB.

The antenna is a 90° corner reflector with two driven dipole elements. Its measured forward gain is 15.5 ÷ 1.0 dB. The estimated E-plane and H-plane half-power beam-widths are 18° and 44°, respectively. The antenna is compact enough to be attached to the cargo hook of any helicopter with high skid gear. On our August 1976 flights we outrigged the antenna from the helicopter in an experimental installaion.

The video output from the receiver modulates the phosphor intensity of a Tektronix model 475 oscilloscope. A slow ramp on the vertical input scans the trace through the full screen height in 1 min. The operator manually controls the vertical scan and a Polaroid-backed oscilloscope camera records the data. This results in approximately 20% of lost coverage by dead time. (In future, we plan to replace the photographic recording system by a magnetic recorder and graphic recorder). (Narod and Clarke, 1980) [28].

Another approach to the sounding of temperate glaciers uses frequencies that are low enough to avoid most of the difficulty arising from scattering. A sounder of this type for surface use has been described by Watts and Isherwood (1978) [46]. The sounder consists of a voltage-step generator based on avalanche-transistors and two identical, resistively-loaded dipoles. The dipoles are loaded with resistors to give a lumped approximation to the relation

$$R(z) = \frac{\Psi}{l - |z|}$$

where Ψ is a constant of proportionality with the ohm as the unit, l is the antenna half-length, z is the distance measured from the center of the antenna, and $R(z)$ is the resistive loading per unit length. Such antennas are very broad-band (Wu and King, 1965) [49], so that, when driven by a voltage pulse, they can radiate a single cycle at a center frequency given by

$$f_c(n) = \frac{50}{l} \text{ MHz}$$

Theoretical and experimental studies of such antennas are considered by Rose and Vickers (1974) [33]; an extension to inductively-loaded antennas, and additional references, are contained in Wright and Prewitt (1975) [48]. Since a center frequency below 5 MHz is desirable in order to avoid large scattering losses, l should be greater than 10 m. Larger antennas also have the advantage of greater effective area both in transmission and reception. The receiver is simply an oscilloscope, sometimes with an additional preamplifier.

In the late 1970s, a low-frequency system specifically designed for the airborne sounding of Columbia Glacier, a highly-crevassed temperate tidewater glacier in Alaska, was developed (Churchill and Wright, 1978) [16]. The system employs a high-voltage, free-running pulser, physically located at the driving point of the transmitting antenna. It is simple, inexpensive, lightweight, and capable of 800 V pulse amplitudes. The pulse time constant is 0.4 μs, and the repetition frequency is ~12 kHz.

In flights over Columbia Glacier, a two-antenna system was employed to avoid the need for a transmit-receive switch, and to allow for different construction of the two antennas while maintaining balance.

The transmitting antenna not only served as a radiator but also provided a high-voltage d.c. path from the power supply on the aircraft to the pulser at the center point of the antenna. The end of the antenna near the aircraft was split into two wires, each of twice the resistive loading per unit length, so that the parallel combination would preserve the correct loading. A half-length of 50 m was used, so that the radiated pulse was approximately a monocycle of 1.5 MHz.

The receiving antenna was similar to the transmitting antenna at its outer end, but the inboard end was a length of coaxial cable over which tubular sections of wire braid connected by resistors were placed. At the center of the antenna the inner conductors of the coaxial cable were attached to the terminals of one side of a balun; the other pair of balun terminals were connected, one to the far end of the antenna, and the other to the outer wire braid of the near end of the antenna.

The antennas themselves were not strong enough to carry stress, so they were limply secured to plastic lines. Plastic funnels 10–15 cm in diameter were used as drogues to hold the antennas out straight and within a few degrees of horizontal

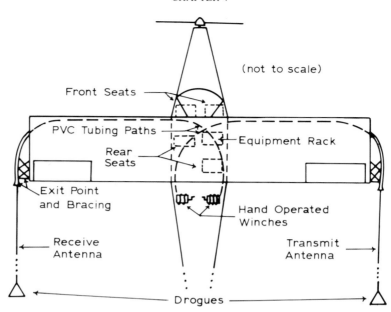

Fig. 29. Antenna deployment and retrieval system, airborne temperate-glacier sounder (Churchill and Wright, 1978).

while in flight. Antennas 50 m long obviously have to be retracted for landings and takeoffs – the method employed involved hand-operated winches inside the fuselage, with the wires fed out through curved pipes fitted to the wing-tips (Figure 29).

4.10. Specialized Equipment

Special equipment has been devised by Walford et al. (1977) [44] for phase-sensitive measurements needed to determine by repeated observations whether the ice thickness in a given location is changing with time (see Section 6.5). The instrument transmits short pulses of radio waves locked in phase with respect to a crystal-controlled oscillator. Radio echoes are received and not only their amplitudes as a function of time, but also their phases with respect to the continuously running oscillator, are observed. Thus all the information carried by the echoes is recorded.

Figure 30(A) shows a block diagram of the radio-echo system. A crystal-controlled oscillator mounted in a thermoelectrically-controlled oven provides a stable reference signal against which phase measurements are made. The oscillator frequency is 59.997 ± 0.001 MHz. The oscillator output drives a timing-control unit which uses emitter-coupled integrated-logic circuitry. Pulses from here synchronize the oscilloscope display and provide a trigger at a pulse-repetition rate of approximately 10 kHz. The trigger switches on a fast thyristor in a pulse modulator circuit which drives the pulse transmitter, a double-beam tetrode of conventional design. The transmitter output is fed through a transmit-receive switch to a wide-band dipole antenna. The amplitude, frequency, and length of the transmitter output

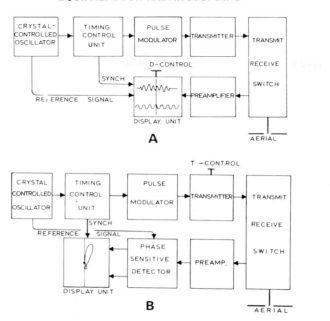

Fig. 30. Block diagram of phase-sensitive radar sounder for measuring change in ice thickness (Walford et al., 1977).

pulse are adjustable. The time at which the transmitter fires, relative to the incoming trigger, can be smoothly, and precisely, varied over a range of approximately 50 ns.

Radio echoes are received and pre-amplified by a low-noise, wide-band, linear amplifier of variable gain. Signals may be displayed in A-scope format, with the reference waveform displayed on a second (lower) trace for comparison purposes. From the display the amplitude of echoes and their phase are measured with respect to the displayed oscillator signal, as functions of the delay time, with the echo sounder at rest on the snow surface.

More precise phase measurement is possible using the phase-sensitive detection system shown in Figure 30(B). From the reference oscillator, two 60 MHz switching signals are derived which differ in phase by $\pi/2$. These switching waveforms and the radio-echo signals are fed to a pair of double-balanced mixers. The action of a double-balanced mixer is to transmit the echo signal unchanged during positive half-cycles of the switching signal and to transmit an inverted version of the echo signal during negative half-cycles of the switching signal. Mixer outputs are smoothed in order to filter out radio frequencies. The resulting video-frequency signals are amplified and fed to the X and Y plates of the oscilloscope, respectively, while the trace is simultaneously brightened. With this system the echo causes a signature to be displayed on the screen that characterizes the amplitude and phase structure of the echo. The instantaneous displacement of the spot from the screen center is proportional to the echo amplitude, the direction of this displacement on the screen is a measure of the echo phase. There is an uncertainty in measuring phase in this way equal to $2\pi n$ radians where n is an integer, but this uncertainty can be removed by referring back to the A-scope display.

The antenna used in experiments on the Devon Island ice cap was a wide-band resonant dipole of adjustable length. It was fed from the transmit-receive switch through a known length of $50\,\Omega$ coaxial cable. A tuned balun was mounted at the antenna's center. Since the admittance of the antenna, and hence the character of the received signal, depend upon the height of the antenna above the surface, that height was kept constant during the experiments. Because the measured phase depends upon the time taken for signals to propagate through the electronic equipment as well as by the propagation time through the ice, observations of the radar echoes were compared with observations made upon a standard echo returned from within a long coaxial cable. Measurements were made by very careful comparisons of the ice-bottom echo with the transmitter pulse and the standard echo, in which instrumental drift was also measured and corrected for. Further details are given by Walford et al. (1977) [44].

References: Chapter 4

[1] Avsyuk, G.A. 'Temperature state of glaciers'. *Trans. Academy of Sciences of the U.S.S.R.*, *Geographical series*, 1955, No. 1, 14–31.
[2] Belousova, I.M., Bogorodsky, V.V., Danilov, O.B., Ivanov, I.P. 'Glacier movement investigations by means of laser'. *Report of the Academy of Sciences of the U.S.S.R.* 1971, **1999** (5), 1055–1057.
[3] Bogorodsky, V.V. *Radio echo sounding of ice*. Leningrad, Hydrometeoizdat, 1975, 63 pp.
[4] Bogorodsky, V.V., Govorukha, L.C., Fedorov, B.A. 'A few results of arctic glacier radio echo sounding'. *Trans. AARI*, 1970, **224**, 87–93.
[5] Bogorodsky, V.V., Trepov, G.V., Fedorov, B.A. 'Variations in the polarization of radar signals in vertical sounding of glaciers'. *J. Technical Physics*, 1976, **46** (2), 366–372.
[6] Bogorodsky, V.V., Trepov, G.V., Fedorov, B.A. 'Possibility of laser use for ice cover dynamic investigations'. *Trans. AARI*, 1970, **295**, 32–34.
[7] Kluga, A.M., Trepov, G.V., Fedorov, B.A., Khokhlov, G.P. 'A few results of summer Antarctic glacier radio echo sounding'. *Trans. XVI SAE*, 1973, **61**, 151–163.
[8] *Handbook to the Catalog of Glaciers of the U.S.S.R.* Ed. by Avsyuk G.A., Leningrad, Hydrometeoizdat, 1966, 76 pp.
[9] Finkelshtein, M.I. *Foundations of Radio Echo Sounding*. Soviet Radio, 1973, 495 pp.
[10] Bailey, J.T., Evans, S. 'Radio echo-sounding on the Brunt Ice Shelf and in Coats Land 1965', *British Antarctic Survey Bull.*, 1968, No. 17, 1–2.
[11] Barnard, H.P. 'Radio-echo sounding in Western Dronning Maud Land, 1974'. *South Africa J. Antarctic Res.*, 1975, No. 5, 37–41.
[12] Barton, D.K. *Radar system analysis*. Prentice-Hall, Microwave and Field Ser, 1964.
[13] Beitzel, J.E. 'Geophysical exploration in Queen Maud Land, Antarctica'. In: A.P. Crary, ed., *Antarctic Snow and Ice Studies II*, American Geophys. Union, Antarctic Res. Ser., 1971, Vol. 16, Washington.
[14] Bentley, C.R., Clough T.W., Jezetz, K.C., Shabtaie, S., 'Ice thickness patterns and the dynamics of the Ross ice shelf'. *J. Glaeiol.*, 1979, **24** (90), 287–294.
[15] Christensen, E.L., Gundestrup, N., Nilsson, E., and Gudmandsen, P., 'Radioglaciology, 60 MHz radar'. Report R-77, Lab. Electromagnetic Theory. Techn. Univ., Denmark, 1970, 82 pp.
[16] Churchill, R.J., Wright, D.L. 'An airborne radio-echo sounder for measurements in temperate glaciers. Final Report'. Colorado State Univ., Fort-Collins, 1978.
[17] Clough, J.W., Bentley, C.R., Poster, C.K. 'Ice thickness investigations on SPQMLT III'. *Antarctic J. U.S.*, 1968, **3** (4), 96–97.

[18] Davis, J.L., Halliday, J.S., Miller, K.J. 'Radio-echo sounding on a valley glacier in East Greenland'. *J. Glaciol.*, 1973, **12** (64), 87–91.
[19] Drewry, D.J. *et al*. 'Airborne radio echo sounding of glaciers in Svalbard'. *Polar Record*, 1980, **20** (126), 261–266.
[20] Evans, S. 'Progress report on radio echo sounding'. *Polar Record*, 1967, **13** (85), 413–420.
[21] Evans, S. 'Radio techniques for measurement of ice thickness'. *Polar Record*, 1963, **11** (73), 406–410; **11** (75), 795.
[22] Evans, S., Smith, B.M.E. 'Radio echo equipment for depth sounding in polar ice sheets'. *J. Sci. Instruments*, 1969, Ser. 2, **2** (2), 131–136.
[23] Evans, S., Robin, G. de Q. 'Glacier depth-sounding from the air'. *Nature*, 1966, **210** (5039), 883–885.
[24] Goodman, R.H. 'Radio echo sounding on temperate glaciers'. *J. Glaciol.*, 1975, **14** (70), 57–69.
[25] Goodman, R.H., Terroux, A.C.D. 'Use of radio echo sounder techniques in the study of glacial hydrology'. *IAHS Pub.*, No. 95, 1973, 149 pp.
[26] Gudmandsen, P. 'Studies of ice by means of radio echo sounding'. In: R.F. Peel *et al.*, eds., *Proc. 28th Symp. Colston Res. Soc., Univ. of Bristal, 1976*, Butterworths, London, 1977, pp. 198–211.
[27] Morgan, V.I., Budd, W.F. 'Radio echo sounding of the Lambert Glacier basin', *J. Glaciol.*, 1975, **15** (73), 103–111.
[28] Narod, B.B., Clarke, G.K.C. 'Airborne UHF radio echo sounding of three Yukon glaciers'. *J. Glaciol.*, 1980, **25** (91), 23–31.
[29] Orheim, O. 'Physical characteristics and life expectancy of tabular icebergs'. *Ann. Glaciol.*, 1980, **1**, 11–18.
[30] Paterson, W.S.B., Koerner, R.M. 'Radio echo sounding on four ice caps in Arctic Canada'. *Arctic*, 1974, **27** (3), 225–233.
[31] Rinker, J.N., Mock, S.J. 'Radar ice thickness profiles – northwest Greenland'. U.S. Army Cold Regions Research and Eng. Lab., 1967, Spec. Rep. No. 103.
[32] Robin, G. de Q., Evans, S., Bailey, J.T., 'Interpretation of radio echo sounding in polar ice sheets'. *Phil. Trans. Roy. Soc. London*, 1969, Ser. A, **265** (116), 437–505.
[33] Rose, G.C., Vickers, R.W. 'Calculated and experimental response of resistivity loaded V antennas to impulsive excitation'. *Int. J. Electron.*, 1974, **37**, 261–271.
[34] Schaefer, T.G. 'Radio echo sounding in western Dronning Maud Land, 1971. *South African J. Antarctic Res.*, 1973, No. 3, 45–52.
[35] Schaefer, T.G. 'Radio echo sounding in western Dronning Maud Land, 1971 – a preview'. *South African J. Antarctic Res.*, 1972, No. 2, 53–56.
[36] Sinsheimer, R.L. 'Altitude determination'. In: *Radar Aids to Navigation*, M.I.T. Radiation Lab. Ser. 2, J.S. Hall, ed., McGraw-Hill, 1947, pp. 131–142.
[37] Smith, B.M.E. 'Airborne radio echo sounding of glaciers in the Antarctic Peninsula'. British Antarctic Survey Sci. Rep., 1972, No. 72.
[38] Strangway, D.W. *et al.* 'Radio-frequency interferometry – a new technique for studying glaciers. *J. Glaciol.*, 1974, **13** (67), 123–132.
[39] Swithinbank, C.W.M. 'Radio echo sounding of Antarctic glaciers from light aircraft'. *IUGG/IASH General Assembly, 1967. Commision of Snow and Ice: Report and Discussions*, IASH Pub. No. 79, 1968, pp. 405–414.
[40] Van Autenboer, T., Decleir, H. 'Airborne radioglaciological investigations during the 1969 Belgian Antarctic Expedition'. *Bull. Soc. Belge Geol., Paleontol. Hydrol.* 1969, **78** (2), 87–100.
[41] Van Zyl, R.B. 'Radio echo sounding in western Dronning Maud Land', 1972. *South African J. Antarctic Res.*, 1973, No. 3, 53–59.
[42] Waite, A.H., Schmidt, S.J. 'Gross errors in height indication from pulsed radar altimeters operating over thick ice and snow'. *Pros. Inst. Radio Eng.*, 1962, **50** (6), 1515–1520.
[43] Walford, M.E.R. 'Radio echo sounding through an ice shelf'. *Nature*, 1964, **204**, 317–319.
[44] Walford, M.E.R., Holdorf, P.C., Oakberg, R.G. 'Phase-sensitive radio-echo sounding at the Devon Island ice cap, Canada'. *J. Glaciol.*, 1977, **18** (79), 217–229.
[45] Walker, J.W., Pearce, D.C., Zanella, A.H. 'Airborne radar sounding of the Greenland ice cap: flight 1'. *Geol. Soc. America Bull.*, 1968, **79**, 1639–1646.
[46] Watts, R.D., Isherwood, W. 'Gravity surveys in glacier covered regions', *Geophysics*, 1978, **43** (4), 819–822.

[47] Weber, J.R., Andrieux, P. 'Radar soundings on the Penny Ice Cap, Baffin Island'. *J. Glaciol.*, 1970, **9** (55), 49–54.
[48] Wright, D.L., Prewitt, J.F. 'Radiating dipole antenna with tapered impedance loading'. *IEEE Trans. Antenna Propagat.*, 1975, **23**, 811–814.
[49] Wu, T.T., King, R.W.P. 'The cylindrical antenna with nonreflecting resistive loading'. *IEEE Trans. Antenna Propagat.*, 1965, **13**, 369–373.
[50] Jiracek, G.R. 'Radio sounding of Antarctic ice'. Geophysical and Polar Research Center, University of Wisconsin-Madison, Research Report 67–1, 1967.

CHAPTER 5

METHODS OF ACQUISITION AND PROCESSING OF DATA

5.1. Radioglaciological Data Recording

In the U.S.S.R. the radar echo sounding of antarctic ice thickness started on a systematic basis on the 11th SAE (1966). A radar of GUYS-1M4 type was used then to measure ice thickness. Recording of thickness measurements was made by photographing amplitude displays ('A-displays') from the oscilloscope screen. The recording interval between frames was 15 s, corresponding to one kilometer of flight distance (in an IL-14 aircraft). Data were recorded both on a photo film and in the flight log. The flight log information included flight direction, flight distance, and the estimated ice thickness. The conditions of recording, shapes and fluctuations of radio signals, operation of individual units, and other useful information for data processing were included in the log. Photofilms and flight logs were stored for possible retrieval and use.

Since 1971 (the 16th SAE), measurements have been recorded photographically using intensity modulation ('Z-display') of the oscilloscope sweep, the film being drawn continuously or in discrete steps across the oscilloscope face, simultaneously with the distance covered. Frame photography continued to be used for special studies. Time marking were made on the continuous recordings from the intensity indicator. This helped to make references to different parts of the flight. Radar sounding results in the Arctic were also recorded, in both A-display and Z-display mode, on photographic film. Flight logs supplement the photo data. Data processing from the films (A-displays) was done visually by counting time delays between echoes, then applying the wave propagation speed through the ice to obtain ice thickness profiles. The Z-display records were also visually analyzed. To help with the analysis the photos were enlarged, using equipment known as a 'Microphot'. The delays between echoes were analysed and ice thickness profiles were plotted.

Computer processing was started in 1974 due to the large volume of information obtained. The processing scheme on the Minsk-32 computer included: film – operator – punch cards – computer. This is not an optimal processing scheme, since it is very time-consuming, particularly at the film–operator stage. However, it gives clear information that is easy to store and to handle (data stored on magnetic tape is easy to retrieve and reproduce if need be).

5.2. Digital and Photographic Recording of Radioglaciological Observations

Bailey and Evans (1968) [1] discuss the importance of photographic recording in connection with work on the Brunt Ice Shelf, Antarctica, using the SPRI Mark II.

They point out that the use of photographic film gives a worthwhile increase in sensitivity by the time integration of the output, obtained from continuous recording, and that the high resolution of transparent-base film is convenient both in storage space and instrumental analysis. These advantages outweigh the difficulties such as the uncertainty, before the film has been developed, that a satisfactory record has been produced, and in some field operations, the difficulty in providing the necessary quantities of clean wash water and suitable drying conditions. It is important in any kind of recording to measure accurately the instant when the transmitter pulse enters the receiver. In the Mark II system this was masked by the initial suppression of the receiver gain; the suppression needed periodocally to be removed so that the direct pulse arrival could be measured. Of course, the travel time of the direct pulse between transmitting and receiving antennas must be allowed for in depth calculations.

Until recently, all radar data were recorded photographically. Amplitude displays (A-displays) were recorded to obtain accurate reflection times to reflectors at base stations, and periodically in the air. Profiles of the ice were made with intensity modulation by sweeping the trace over the plane of the film while towing the entire system over the ice. With pulse repetition frequencies near 10 kHz, substantial signal averaging occurs on the oscilloscope screen and film. This eliminates many spurious effects and enhances the continuity of real features. Thus, profiles yielded a good understanding of the relationships between reflectors at different horizontal positions. From radar profiles, evidence for features such as internal layers (continuous for tens of kilometers), bottom crevasses, buried surface crevasses, and subglacial lakes has been found.

Photographic recording techniques suffer from several disadvantages, however, which become more pronounced when the focus of the investigation shifts toward finer details within the ice. Most of these difficulties are associated with the profile method of data collection. Profiles are adequate for most qualitative investigations, but less suitable for quantitative studies. Accuracies in picking travel times are inherently less than those obtained from A-displays. Quantifying profiles from high-density travel time measurements is tedious at best. The combination of oscilloscope intensity-modulation circuitry and film characteristics limit the dynamic range of profiles to about 10 dB. However, the system response (the attenuation required to drown the transmitted pulse in noise when the transmitted pulse is fed directly into the receiver) is about 160 dB for most radars. Finally, the nature of profiles implies that all amplitude and phase information is lost.

Many of the problems associated with profiles could be solved by recording A-displays, but several complications limit this. It is almost impossible to correlate reflections from horizontal strata at different A-display recording sites. Difficulties are caused by confusing diffraction echoes which result from changes in the topographic and reflection characteristics of the layers. It is possible to find the general locations of the reflecting horizons only by carefully plotting the times measured from many A-displays along a traverse line. Even if echoes could be easily correlated, immense quantities of film would be required to produce records comparable in quality to profiles currently being produced. Since film records are very difficult to reduce to numbers, this approach is generally unfeasible. An early attempt at digital recording was carried out on the U.S. Queen Maud Land traverse in 1968–69 (Beitzel, 1971) [3]. The system was designed to record the vertical travel times at selected intervals and store them on paper tape, as well as to plot a chart of the ice thickness. Depths up to 3500 m were routinely measured with this system. Unfortunately, because of the termination of the U.S. oversnow traverse program, this capability was never pursued.

A more modern digital recording system was used by Goodman (1975, Chapter 4, [24]). In his system the radar echo is digitized using a sampling technique and the resulting histogram is stored in the computer. The computer controls the time delay between the transmit pulse and the sampling of the data. The total time delay to the recorder is thus divided into 1024 time channels, only one of which is sampled per radar return. After 1024 returns, the total histogram has been constructed. This process takes about 20 ms and the time required to write it on magnetic tape is about 1 s. From time considerations, up to 50 events could be averaged, but the computer work length limits averaging to 16 events per record. The difference in the statistical error between an average of 50 events and 16 events is less than a factor of two. A display reconstructed from the histogram allows the operator to monitor the quality of the data.

Valuable information would be available if, in addition to accurate travel times, amplitudes and phase data could be obtained from return echoes. A recording scheme to do this has been developed at the University of Wisconsin-Madison. Since any new technique would be required to handle greatly increased data loads, digital processing with magnetic tape storage was chosen as the most practical system.

The digital recording system (Figure 31) is built around a SPRI Mark IV radio echo sounder modified to operate at a center frequency of 50 MHz. From the receiver the signal is sent down parallel lines terminating in the recording oscillo-

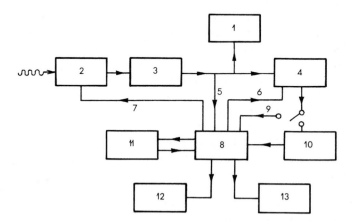

Fig. 31. Block diagram of University of Wisconsin digital recording system. 1 – recording oscillograph; 2 – time-varying gain; 3 – receiver; 4 – transient recorder; 5 – trigger; 6 – start scan; 7 – test signal; 8 – computer; 9 – phase data; 10 – signal averaging (done by computer); 11 – digital tape recorder; 12 – oscilloscope; 13 – graphic recorder.

graph and the digital recording system. The oscillograph is used as a real-time monitor and produces profiles. The heart of the digital system is the high-speed transient recorder, a device to sample and store in memory a single event at sampling rates from 0.1 Hz to 100 MHz with a total memory length of 2048 8-bit words. The 8 bits yield a dynamic range of 48 dB. Coupled with time-varying gain, this range should be about 30 dB better than that of amplitude displays recorded on film. The transient recorder has none of the signal continuity problems associated with sample-and-hold techniques. This means that changes in velocity of the recording system will only affect the horizontal sampling of scans, not signal reconstruction.

A minicomputer (PDP 11/03) controls the entire system, taking data from the transient recorder to be stored on 9-track magnetic tape. It also processes data for output visually on an oscilloscope and on a graphic recorder (dynamic range 23 dB) for profile reconstruction. When studying the deepest layers, noise levels can be reduced by signal averaging in the computer. The system can record at a maximum rate of 3 Hz. A single tape, 18 cm in diameter, holds more than 2000 records. The 3 Hz recording rate was designed to ensure that the density of sample points along a surface traverse is satisfactory.

The concept of this system offers many advantages over past techniques. Since all information detected by the receiver is stored, profiles can be constructed under much greater control. Rather than spending large amounts of field time obtaining profiles over the same line with varying intensities, horizontal sweep rates and attenuation, the same types of reconstruction can be made in the laboratory. It should also be possible to enhance profiles constructed from digital signals by filtering, migration, and magnification of the profile. Data used to pick travel times can be used to construct profiles – ensuring accuracy and the ability to correlate reflectors. Finally, the increased dynamic range should allow the study of deeper and fainter features in the ice than has previously been possible.

The digital recording system, built in 1978, was tested at Dome C during the 1978–79 field season (Bentley et al., 1979) [2] and found to require considerable modification. Most importantly, it was found that the interface between the computer and the tape recorder radiated unacceptable amounts of radio-frequency noise. Consequently, the system has been modified by removing the computer from the real-time system; it instead will be kept in the base camp for use at the end of the day. The new vehicle-mounted system consists of the transient recorder, a microcomputer, a memory add-on circuit, and a tape recorder.

Because the new system hardware is designed to do specific tasks, it is much faster than the old system. As a result, it is possible to do fast, real-time averaging before writing on tape. Approximately 200 records from the transient recorder can be stacked in the added memory and only the stacked signal is recorded on tape, at about 4 records per second (the number of records stacked is selectable – 200 records is a maximum). Real-time signal averaging represents a considerable increase in the amount of information being recorded. The old design recorded a signal every 0.3 s. Since the transmitter can fire at a 10 kHz rate, only one record in 3000 was being saved. The new system saves almost one signal out of ten. Furthermore, the signal averaging should improve the signal-to-noise ratio by about 25 dB, effectively increasing the system response of the SPRI Mark IV 50 MHz radar to 175 dB.

References: Chapter 5

[1] Bailey, J.T., and Evans, S. 'Radio echo sounding on the Brunt ice shelf and in Coats Land, 1965'. *British Antarctic Survey Bull.*, 1968, No. 17, 1–12.
[2] Bentley, C.R. *et al.* 'Geophysical investigation of the Dome C area'. *Antarctic J. U.S.*, 1979, **XVI** (5), 98–100.
[3] Beitzel, J.E. 'Geophysical exploration in Queen Maud Land, Antarctica'. In: A.P. Crary, ed., *Antarctic Snow and Ice Studies II*, American Geophys. Union, Antarctic Res. Series, 1971, Vol. 16, Washington.

CHAPTER 6

SCIENTIFIC RESULTS IN RADIOGLACIOLOGY

6.1. Direct Measurement of Permittivity and Conductivity

LABORATORY MEASUREMENTS

In Chapter 3, we discussed the principal measurements on the electrical properties of ice grown in the laboratory and our discussion touched briefly on the electrical properties across the entire frequency spectrum. We will now consider the properties of natural glacier ice and, in doing so, concentrate our attention on the range of frequencies that are of concern in radioglaciology, i.e., 10^4–10^9 Hz. In this range, both ε' and σ_∞ are essentially constant as a function of frequency. ε' is also nearly constant with temperature, whereas σ_∞ varies according to an activation energy that is essentially the same as that for the characteristic relaxation time, τ. In this part of the spectrum the Debye theory works very nicely, so that we infer results about the temperature dependence of τ from measurements of σ_∞ – most of the actual measurements are of the latter quantity, which is directly involved in the energy absorption rate in the ice.

MONOCRYSTALLINE ICE

The only naturally occurring monocrystalline ice upon which measurements have been made comes from Mendenhall Glacier. Several sets of measurements have been made on the large single crystals of Mendenhall Glacier, the most complete set being those of Paren (1973, 1970) [137, 201]. At temperatures above $-60°C$, this ice shows Debye relaxation characteristics that are very similar to those of pure ice. The ice is apparently so pure that its dispersion characteristics are substantially closer to the ideal than are those exhibited by most laboratory-grown ice (Glen & Paren, 1975) [82]. The differences between various ice specimens become apparent in a plot of high-frequency conductivity against reciprocal temperature (Figure 32), where it can be seen that the Mendenhall crystals continue along the Arrhenius line to lower temperatures than other natural samples. Ackley and Itagaki (1969) [24] measured the effect of uniaxial compressional strain on Mendenhall ice. They found that the imaginary part of the permittivity was changed by more than 10% in the whole Debye dispersion range for frequencies less than 50 KHz, but again no significant effect was shown at MHz frequencies. The measurements on σ_∞ reported for Mendenhall Glacier were made with the electric field parallel to the c-axis.

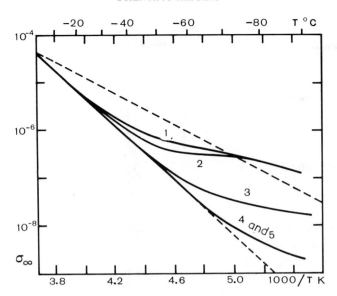

Fig. 32. Experimental determinations of the temperature dependence of the high-frequency conductivity. Dashed lines show slopes for activation energies of 0.26 eV (upper line) and 0.56 eV (lower line), respectively. Experimental curves are: 1 – ice from TUTO tunnel, Greenland, taken 100 m from the portal (Paren, 1970). 100 kHz; 2 – polycrystalline ice (Camp et al., 1969). 20 kHz; 3 – single crystal, electric field $\perp$ c-axis (R. Ruepp. quoted in Glen and Paren, 1975). 300 kHz; 4 – single crystal, electric field $\parallel$ c-axis (R. Ruepp, quoted in Glen and Paren, 1975). 300 kHz; 5 – Mendenhall Glacier single crystal, electric field $\parallel$ c-axis (Glen and Paren, 1975). Frequency varies with temperature, from about 50 kHz at 0°C to 100 Hz at −90°C. Curves 4 and 5 coincide.

POLYCRYSTALLINE ICE

Whereas the only single-crystal ice was from a temperate glacier, all the measurements on polycrystalline ice have been made on ice from the polar ice sheets. These results are, of course, the ones that are most directly applicable to radar studies. It is widely assumed that the value of ε_∞ does not vary significantly from place to place on polar ice, except for a small temperature coefficient.

A comprehensive set of measurements taken by Westphal was summarized by Jiracek (1967) [110]. For ice of density 0.90 mg m^{-3} from Greenland and the Ward Hunt Ice Shelf, ε' ranged from 3.20 at −1°C to 3.13 at −60°, with no variation over the frequency range 150–2700 MHz. Corrected to $\rho = 0.917$ mg m^{-3} according to Looyenga's equation, this range becomes 3.26 to 3.19. Less dense ice from Little America Station on the Ross Ice Shelf ($\rho = 0.881$ mg m^{-3}), and an unspecified arctic glacier ($\rho = 0.835$ mg m^{-3}) gave 3.07–3.00 and 2.88–2.83, respectively, over the same temperature range. Correction for density yields ranges of 3.19–3.11 and 3.14–3.08, respectively, for the Ross ice shelf and the arctic glacier. These lesser values (compared to Greenland and the Ward Hunt ice shelf) could reflect an inaccuracy in the correction for density, perhaps due to variations in the Formzahl, since it is not known how the samples were oriented relative to the electric field, or they may reveal real differences in ε_∞ in polar ice. Low values of ε'_∞ have been suggested by other laboratory measurements (next paragraph) and by some field studies.

The best recent measurements have been made on ice from the deep core at Byrd Station in West Antarctica (Fitzgerald and Paren, 1975) [79]. Measurements were made on two different samples, one from a depth of about 150 m, where the crystalline fabric is nearly isotropic, and the other from a region around 1400 m, in which recrystallization has given rise to a very strong vertical preferred orientation parallel to the c-axes. The dielectric response was studied between 60 Hz and 10 kHz, with enough measurements at higher frequencies to be confident of the high frequency extrapolation. Temperatures were varied between $-6°$C and $-60°$C. Despite the strikingly different fabrics in these two samples, no significant electrical differences were found. At $-60°$C, $\varepsilon_\infty = 3.13 \pm 0.02$. However, when the sample was melted and refrozen, ε_∞ increased to 3.18 ± 0.01. This suggests that the metamorphic ice does show a smaller value of ε_∞ than ice formed by freezing from a melt. A still more striking result was reported by Paren and Glen (1978) [138] from measurements on another sample taken at the same depth from the drill hole at Byrd Station as Fitzgerald and Paren's deep sample. Their experiment yielded a value of ε_∞ of only 2.9 at $-45°$C. The latter result is particularly interesting in the light of measurements of ε_∞ measured *in situ* in the field as discussed below.

The differences between values of σ_∞ for the Byrd core and for pure ice (or Mendenhall Glacier ice) are large. As summarized by Glen and Paren (1975) [82] it has been known for some time that σ_∞ for deep polar ice exhibits an activation energy of only about 0.25 eV compared with 0.57 eV for Mendenhall Glacier ice and pure laboratory ice. Again, the measurements from the Byrd Station deep core provide good evidence. Measurements have been made both by Fitzgerald and Paren (1975) [79] and by Maeno (1974) [126]. As shown in Figure 33, the two sets of measurements are in excellent agreement with each other and also with measure-

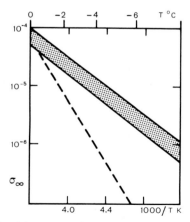

Fig. 33. High-frequency conductivity measured on samples of ice from the Byrd Station deep core, Antarctica. Shaded area shows range of measurements by Maeno (1974) and Fitzgerald and Paren (1975). Dashed line shows variation in pure ice.

ments on ice from Camp Century and site 2 in Greenland as studied by Paren (1973) [137]. The striking difference in activation energy shows clearly in Figure 33. Since the deep polar core ice, with a very small bubble volume, behaves just the same as more bubbly ice from nearer the surface, the difference must arise from the bulk property of the ice. A second interesting feature of the measurements is that, at 0°C, the values for pure ice (and Mendenhall Glacier ice) and polar ice are nearly the same: $\sigma_\infty = 5 \times 10^{-5}\,\Omega^{-1}\,m^{-1}$. Other similar results have come from ice samples from the Ward Hunt ice shelf in Canada (Westphal, quoted in Evans, 1965) [75], and the Mizuho Plateau in East Antarctica (Maeno, 1974) [126]. Two sets of measurements by Westphal (quoted in Glen and Paren, 1975) [82] on ice from an unidentified location in the Arctic and from Little America on the Ross ice shelf in Antarctica, show similar values of σ_∞ at T = 0°C, but an activation energy of only 0.17 eV.

An exception to this uniformity in value of σ_∞ at 0°C in polar ice would seem to be found in some cores of lower density, corresponding to ice that is permeable (Maeno, 1974 [126]; Paren, 1973 [137]) where higher conductivities are reported. These data need to be corrected to allow for the presence of air inclusions, as will be discussed below for the case of snow, but this further increases the conductivity. The activation energy of σ_∞ is, however, similar to that of polar ice, and the phenomenon could be due to contamination of the permeable core during the drilling process.

Another deviation from the general pattern was reported by Paren (1970) [201] on ice from the Tuto Tunnel that showed a change in activation energy at about −30°C from 0.57 eV to 0.15 eV (Figure 34). The net result over the temperature range from 0 to −80°C is about the same as an average activation energy of 0.25 eV.

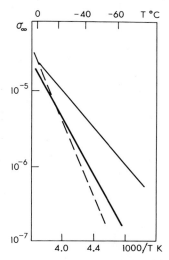

Fig. 34. High-frequency conductivity measured on melted and refrozen ice sample from the Byrd Station deep core, Antarctica (heavy line). Lighter line is σ_∞ in unmelted core, from Figure 33. Dashed line shows variation in pure ice. From Fitzgerald and Paren (1975).

In the melted and refrozen Byrd core sample (Fitzgerald and Paren, 1975) [79] the results obtained were substantially different from those of the original samples (Figure 34). The activation energy was increased to reach a value close to that for pure ice, although it remained slightly lower. In fact, all the melted and refrozen samples behave like ice very lightly doped with HF. According to Camplin and Glen (1973) [47], the concentration of HF required to produce the same effect would be less than 5×10^{-5} mol l^{-1}.

A few measurements of the characteristic relaxation frequency have been given. Paren (1973) [137] gives $f_c = 0.7$ kHz at $-45°$C from measurements on ice from Camp Century, Greenland. Subsequent measurements by Paren and Glen (1978) [139] on ice from the same location gave $f_c = 0.6$ kHz at the same temperature – activation energy is not given in this work. Paren (1973) [137] quotes an activation energy of only 0.16 eV for the Camp Century ice.

For the determination of depth from radar sounding, the variations in ε_∞ reported are small enough to produce no significant uncertainty in ice thickness, although the deviations are of interest in terms of the physical characteristics of the ice itself. On the other hand, when efforts are made to use absorption measurements to determine temperature, or to calculate absorption for use in studies of layer reflectivity, more care needs to be taken. Despite the preponderance of measurements that show close inter-agreement for σ_∞ and its activation energy, uncertainties about the effect of impurities, and particularly the effect of annealing of ice involving some melting and refreezing, require that great caution be employed. Nevertheless, the general agreement between the various measurements of σ_∞ does hold out substantial hope for the use of absorption measurements in radioglaciology.

We have been concerned so far solely with the principal Debye dispersion in ice. Careful laboratory measurements on both laboratory ice and polar glacier ice does show that the dispersion in ice does not exactly follow the simple Debye law, but that there are deviations which can be analyzed as one, or several, other small dispersions. These additional dispersions are best shown by the work of Von Hippel *et al.* (1972) [176]. At the present time these additional dispersions, some of which occur at frequencies higher than the principal dispersion, are both too small and too poorly specified to be of concern in remote sensing. However, it is possible that additional refinement of the experimental evidence will in the future make it possible to take the high-frequency dispersions profitably into account.

SNOW

There is a decided paucity of measurements on snow, particularly on samples with a density such as is found in firn on [27] glaciers. In fact, only one set of measurements exists for densities higher than 0.43 mg m^{-3}. Nevertheless, adequate corrections for firn layers on polar ice caps probably can be made.

Measurements of ε' on natural snow samples include principally those by Cumming (1952) [61] at 9 GHz on Ottawa snow, and Kuroiwa (1956) [122] at

3 MHz on snow from Sapporo, Japan. The combined results are shown in Figure 35, together with lines representing Weiner's equation with $u = 2$ and $u = \infty$, and Looyenga's equation. It can be seen that either Looyenga's equation or Weiner's equation with $u \simeq 5$ fit the data rather well. For this reason, correction for ε' in the polar firn layers can be made with some confidence if the density is known as a function of depth.

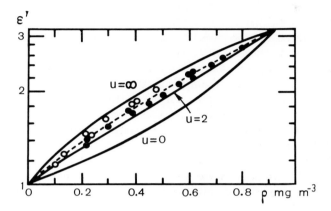

Fig. 35. Variation of the real part of the dielectric constant, ε', with density, ρ (mg m^{-3}), as measured by Cumming (1952) (solid circles), and Kuroiwa (1956) (open circles), Solid lines are from Weiner's equation; dashed line is from Looyenga's equation. From Evans, 1965.

Cumming (1952) [61] also measured ε'', finding values that agreed well, for densities between 0.3 and 0.8 mg m^{-3}, with Weiner's equation using $u = 2$. Unfortunately, there seem to be no other measurements for densities greater than 0.4 mg m^{-3}, about the minimum for polar firn.

Kuroiwa (1962) [123] did measure σ_∞ as a function of frequency for low-density snow, as did Keeler (1969) [114] on snow from Alta, Utah, U.S.A. and Ambach and Denoth (1972) [27] on snow from the Austrian Alps. It is generally found that there is a prominent dispersion in snow in the same frequency range in which the Debye dispersion occurs in ice. However, the characteristic relaxation frequency is some five times higher than that to be expected from the value for ice, and the variation in the complex permittivity as a function of frequency is substantially different. The high-frequency conductivity of the ice component deduced from data on snow covers a two-order-of-magnitude range at temperatures within 10°C of the melting point; σ_∞ ranges from 5×10^{-6} to $5 \times 10^{-4} \, \Omega^{-1} \, \text{m}^{-1}$. Fujino (1967) [80] measured the temperature dependence of the relaxation frequency of one snow block of density 0.38 mg m^{-3}, finding an activation energy of 0.27 eV, very similar to that for σ_∞ in polar ice.

Traub and Gribbon (1978) [171] measured activation energies for low-density temperate snow samples collected in Scotland, finding that values were a factor of two greater at temperatures above $-25°C$ than at colder temperatures. For dry samples, the values were 0.42 eV and 0.18 eV and, for both wet and refrozen snow, 0.60 eV and 0.26 eV, respectively. High-frequency conductivities calculated for the different snow samples range from 5 to $13 \times 10^{-15} \Omega^{-1} m^{-1}$ with no particular pattern evident, and with an average about twice that found by direct measurements on solid ice. On the other hand, if the measurements for low temperature are extrapolated to $0°C$, the value of σ_∞ indicated for ice is about half that for direct measurement.

The best study of the problem has been made by Paren and Glen (1978) [139] on artificial ice produced by shaving solid ice with a cheese grater. 'Snow' of density ~ 0.42 mg m^{-3} was produced both from pure single crystals of ice and from a sample taken from a depth of 1424 m in the Byrd Station deep borehole. No significant differences were found between the samples thus produced, despite the large differences between pure ice and polar glacier ice that have already been described. Furthermore, the activation energy took on the same value as for polar ice: ~ 0.25 eV. The relaxation frequency extrapolated to $0°C$ was ~ 30 kHz, in good agreement with Fujino's value.

Paren and Glen (1978) [139] found their measurements to be interpreted best by two dispersions: the main relaxation spectrum and a high frequency dispersion with f_c about 10 times as great, and $\Delta\varepsilon$ about 1/10 as large, as for the principal dispersion. Since the contribution of a particular dispersion to σ_∞ is proportional to the product of f_c and $\Delta\varepsilon''$, $\Delta\sigma_\infty$ for the two was very nearly the same. (For the Byrd borehole ice itself, the principal contribution to $\Delta\sigma_\infty$ is about four times as large as the high-frequency value.)

The solid-ice Byrd core sample was analyzed in the same way, and the comparison with extrapolated ice-component values from the snow samples yielded strong differences. Although the values of σ_∞ were not too different, the characteristic frequencies for both the principal and the high-frequency dispersions and, consequently, the corresponding values of σ_∞, were ten times higher as calculated for the ice-component values using Looyenga's equation as for the parent ice sample. In fact, f_c for the high frequency dispersion in the Byrd core ice sample is essentially the same as that for the principal dispersion in the finely-divided ice samples. The reasons for this discrepancy are not clear – the result is in striking contrast to measurement *in situ* in the Canadian Arctic and Antarctica (see the next section).

FIELD MEASUREMENTS

Temperate Glaciers

There have been surprisingly few measurements of the dielectric dispersion of temperate glacier ice *in situ*. The only measurements to cover the main Debye dispersion with a reasonably accurate technique are those of Watt and Maxwell (1960) [185] on Athabasca Glacier, whose temperature characteristics have been studied in detail by Paterson (1971, 1972) [140, 141]. They used both two-terminal bridge measurements and four-terminal a.c. conductivity measurements, the results of which were consistent and gave a Debye relaxation frequency which is completely consistent with that found for laboratory polycrystalline ice, though they found considerable distortion of the simple Debye dispersion at low frequencies, probably associated with a high d.c. conductivity.

Radio-frequency interferometry can be used as a method of measurement of the dielectric properties of ice *in situ*. Waves propagate from a transmitter on the ice surface to a separated receiver along paths just above and just below the surface. Since the two waves travel at different speeds, an interference pattern that depends upon the dielectric constant of the ice develops (Hermance, 1970) [104]. Application of this method to measurements on Athabasca Glacier, using frequencies from 1 to 10 MHz, led to reasonable estimates of both ε' (slightly less than 3.2) and σ_∞ ($\approx 10^{-4}\,\Omega^{-1}\,m^{-1}$) (Hermance and Strangway, 1971 [105]; Strangway *et al.*, 1974 [166]), but the accuracy of the method is not sufficiently high to yield a definitive determination of these quantities.

Polar Glaciers

Most of the field experiments relating to the determination of ε' on polar ice bodies have been designed to measure the wave propagation velocity directly, thus yielding ε'. Those measurements are described in a later section. Here we will summarize only the meagre experimental field evidence relating to the dielectric properties near the main relaxation spectrum.

Rogers and Peden (1973) [156] have measured the VLF electrical properties of the ice sheet *in situ* by lowering a 3 m dipole probe into the deep drill hole at Byrd Station and measuring probe admittances at five frequencies between 1.25 and 20 kHz. They found results for the complex permittivity which they interpret as being in close agreement with those for pure ice at the same temperature, and therefore at variance with measurements made on polar glacier ice. At frequencies less than 5 kHz, the measured values of ε' and ε'' fell increasingly below those for pure ice as the frequency was decreased, thus increasing the discrepancy relative to 'normal' polar ice values. Fitzgerald and Paren (1975) [79] suggest that the discrepancy may be due to contamination of the ice by the drilling fluid, both through its electrical effects as a proton acceptor and through the melting and refreezing of ice in the immediate vicinity of the hole.

Effective average values of the complex dielectric constant in the Antarctic ice near Byrd Station have been calculated using measurements of the phase and amplitude of VLF emissions from the 34 km long antenna near Byrd Station (Peden *et al.*, 1972) [145]. The results of these studies at frequencies between 5 and 20 kHz can be fitted fairly well with a section of a semi-circle on a Cole–Cole plot, i.e., by assuming a simple relaxation spectrum that corresponds to a single relaxation frequency of about 2.5 kHz. This would correspond to an effective mean temperature in pure ice of about $-10°C$, or an effective mean temperature in polar ice of about $-32°C$. Peden *et al.* (1972) [145] point out that $-10°C$ is a reasonable effective mean temperature in view of the exponential increase in absorption to be expected with increasing temperature in the lower part of the ice sheet, and accept the pure-ice model. However, this again suggests a glaring discrepancy with the other measurement of the dielectric properties of polar ice. Glen and Paren (1975) [82] point out that σ_∞, as measured by Peden et al. (1972) [145] is nearly constant at lower frequencies, then decreases at higher frequencies, a behavior directly the opposite of that to be expected for a Debye dispersion. They suggest that the different sampling depths in the earth (all much larger than the thickness of the ice sheet) at different frequencies may be responsible.

Another measurement of the bulk dielectric loss in the Antarctic ice sheet comes from a study of attenuation in long-range VLF transmission along paths that cross the Antarctic ice sheet (Crary and Crombie, 1972) [59]. Large excess attenuation rates found over these paths are attributed to losses in the Antarctic ice, leading to an estimate of the dielectric constant, $|\varepsilon|$, at a frequency of 16 kHz as lying in the range 10–50. This is not a closely defined value, but the range is correct compared with the measurements of Fitzgerald and Paren (1975) [79], and it is an interesting approach to an average dielectric constant over a long path. Perhaps the method can be refined in the future.

Snow

Contrary to the situation with regard to laboratory measurements on snow, *in situ* observations have yielded results that agree well with the bulk properties of polar ice.

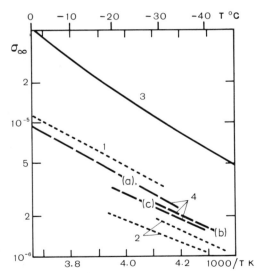

Fig. 36. Variation of the high-frequency conductivity measured *in situ* in firn on White Glacier, Axel Heiberg Island at 20 kHz (upper dashed curve (short dashes); Glen and Paren, 1975) and Byrd Station, Antarctica at 12.8 kHz (lower two short-dashed curves; Webber and Peden, 1970; Peden and Rogers, 1971). Solid line shows measurements on solid ice from Greenland by Paren (1973), from which the dashed curves (long dashes) were calculated (upper curve; density 0.43 mg m^{-3}, 20 kHz; middle curve: density 0.40 mg m^{-3}, 12.8 kHz; lower curve: density 0.40 mg m^{-3}, 10 kHz). From Glen and Paren (1975).

Measurements on polar snow *in situ* have been made by Paren (1970) on Axel Heiberg Island, Canada, and by Rogers (unpublished, reported by Webber and Peden, 1970 [188]; Peden and Rogers, 1971 [144]) at Byrd Station, Antarctica. The results are complementary, in that Paren's data cover temperatures between 0°C and −30°C, and Rogers' data cover temperatures between −19°C and −44°C. The frequencies in both cases (Peden: 20 kHz max; Rogers: 13 kHz max) were high enough to give good determination of σ_∞. The results are shown in Figure 36, along with calcualted curves based on measurements on deep Greenland ice and mixture theory. The agreement is good; note that no such agreement would have been obtained had bulk values for pure ice rather than polar (Greenland) ice been used.

6.2. Electromagnetic Wave Speed and its Measurement

The electromagnetic wave speed in ice in one of the most important glacier parameters; it determines the accuracy of glacier thickness measurements and indicates variations in ice density. There are basically two different ways of determining the speed. The first is [12] to use laboratory measurements of the relative permittivity of ice as a function of temperature and density, and then, using (1.3), apply those measurements to the glaciers and ice sheets where temperatures and densities are known. There are several problems inherent in this method, however. Temperatures and densities at depth in thick ice bodies are often not very well known. Furthermore, there may be other influences on the wave speed, such as a preferred crystal orientation and impurities in the ice; even the strain history of the glacier ice might have an effect. Thus, for direct application to ice thickness determinations, it is essential to have measurements of the electromagnetic wave speed *in situ* as well as in the laboratory. Such measurements are designed to measure the wave speed directly, from which the permittivity can be calculated if desired, rather

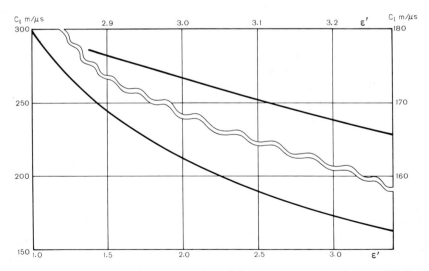

Fig. 37. Plot of electromagnetic wave speed vs. dielectric constant, ε', using $c = 299.8 \, \text{m} \, \mu\text{s}^{-1}$.

than proceeding in the reverse direction. For convenience, a plot of velocity vs. relative permittivity is given in Figure 37.

There are several different geophysical techniques that can be used for such *in situ* determinations of wave speed. These include: logging in drill holes, sounding next to drill holes, studies of propagation between drill holes, oblique reflection sounding, and comparisons with seismic reflection soundings. We will discuss these in turn.

LOGGING IN DRILL HOLES

The most precise method of determining velocities, and the one that also gives the most detail in the variation of velocity with depth, is the interferometry technique developed by Robin and applied in a drill hole on the Devon Island ice cap in June 1973 (Robin, 1975) [151]. The principle of the technique is to lower down the bore hole an antenna which radiates a continuous signal at a fixed frequency, to pick up the radiated signal after it has passed through the ice and firn to a second antenna on the surface, and then to mix the received signal with the original fixed frequency to generate 'beats' as the antenna is lowered or raised. By measuring the spacing between successive null points the velocity of propagation can be determined by multiplying wavelength by frequency.

Measurements by this technique on Devon Island were accurate only when the down-hole antenna was below a critical depth of about 50 m. At shallower depths there are difficulties with the pulse shape, having to do with energy reaching the receiving antenna by different paths, which makes measurement of null points impracticable. At greater depths the internal consistency of the measurements, determined by comparing values from upward and downward runs of the antenna, was within $\pm 1 \, \text{m} \, \mu\text{s}^{-1}$. A cyclic variation with depth of several meters per microsecond amplitude was found, but it differed in phase relative to depth upon separate runs in the hole, and was therefore attributed to some sort of wave propagation phenomenon.

Robin's (1975) [151] mean experimental value for the wave velocity in the ice cap at a mean $T = -20°C$ and a mean frequency of 425 Hz, averaged over the apparent depth cycle, is $168.6 \pm 0.2 \, \text{m} \, \mu\text{s}^{-1}$, corresponding to $167.5 \pm 0.2 \, \text{m} \, \mu\text{s}^{-1}$ ($\varepsilon' = 3.21 \pm 0.01$) after correction to $\rho = 0.917 \, \text{mg m}^{-3}$. This value should be considered the best *in situ* determination of c_1 that has been made to date.

SOUNDING NEXT TO DRILL HOLES

Another potentially accurate way of determining the mean velocity through the ice sheet is to carry out reflection sounding next to points where the thickness of the ice is accurately known by actual drilling down to the bed. However, the absence of echoes precisely at the drill hole site have so far led to large interpolation errors that have prevented accurate velocity calculations on thick inland ice. The first measurements of this kind were carried out next to the first drill hole through a major ice sheet, that at Camp Century in Greenland. Ice soundings by Pearce and Walker (1967) [143] yielded a contour map of the two-way reflection time over a region about 5 km on a side centered on the drill hole. The interpolated time at the drill hole site was 16.75 μs. The total depth of the drill hole was 1391 m leading, after an 8 m correction for the firn layers, to a velocity in the solid glacier ice of only 165 m μs^{-1} ($\varepsilon' = 3.30 \pm 0.04$). This is a very low value that has not been reproduced elsewhere from field measurements.

Independent radar reflection measurements in the vicinity of the Camp Century drill hole were made by Robin et al. (1969) [153] and yielded reflection times within 300 m of the drill hole of 16.2 and 16.0 μs in two passages of the snow vehicle. After an 8 m correction for the firn layers, these times yield a velocity of 171.5 $\pm$ 1.5 m μs^{-1} ($\varepsilon' = 3.07$), much higher than Pearce and Walker's.

The difference between the two sets of measurements at Camp Century cannot easily be explained. Since Pearce and Walker had only one reflection point within 1 km of the drill hole, there could be an interpolation error of one or two tenths of a microsecond, but it seems highly unlikely that Pearce and Walker's reflection points do in reality define a surface lying 50 m below the base of the drill hole. In view of the fact that higher velocities are found virtually everywhere else, we can only treat their result as an unexplained anomaly.

In 1967–68, soundings were made from an aircraft in the neighborhood of the 2164 m drill hole at Byrd Station. Ice thicknesses of between 2104 and 2214 m were obtained using $c_i = 169$ m μs^{-1} (Robin et al., 1970) [155]. However, this cannot be considered useful as a velocity determination – inverting their calculation, we find only that the velocity lies between 165 and 174 m μs^{-1}. From a careful re-examination of the same records, Drewry (1975) [69] found weak, but distinguishable, echoes at a distance of about 400 m from the hole, from which he calculated a thickness of 2150 $\pm$30 m, also using $c_i = 169$ m μs^{-1}. Accepting his error figure, and inverting, yields a reduced velocity range of 170 $\pm$2.5 m μs^{-1}.

In 1979 electromagnetic wave speed measurements were made on Vavilov Glacier (Severnaya Zemlya) (Fedorov, 1978 [23]). The thickness of the glacier was measured next to two drill holes located 65 m apart, and by a third drill hole in the southeast part of the glacier, 11 km away from the first two. Values of electromagnetic wave speed determined from these soundings are given in Table VIII.

TABLE VIII
Electromagnetic wave speed measured at Vavilov Glacier

Hole number	Glacier thickness in the hole (m)	Signal propagation time (μs)	Mean speed (m μs^{-1})
1	458	5.5	165.5
2	462	5.5	168.0
3	551	6.6	167.0

OBLIQUE REFLECTION SOUNDING

The second most precise method of determining electromagnetic wave velocity in ice sheets and glaciers is oblique reflection sounding. This method was successfully

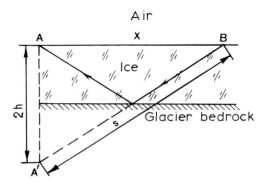

Fig. 38. Schematic graph of wave velocity measurement by oblique-reflection technique.

used in Greenland and Antarctica. This is a standard technique of exploration geophysics, commonly used for seismic measurements, in which the variation of the travel time for a bottom reflection is measured as the transmitter and receiver are separated along the surface.

Figure 38 illustrates the essence of oblique reflection sounding. The transmitter and pulse oscillator are located at point B, and the receiving antenna with receiver and oscilloscope are at A, a distance X away. Two pulses are recorded: the direct pulse, which has gone a distance X through the air, and the pulse reflected from the bedrock which has traversed-length S. Point A' is the image of point A, relative to the bedrock boundary. By measuring the time interval between the reflected and direct signals, it is possible to determine the electromagnetic wave velocity in ice. If the assumption is made initially that the ice is homogeneous and isotropic and that the top and bottom surfaces of the ice are parallel, then simple geometry leads to a relationship:

$$c_i^2 t_s^2 = x^2 + 4h^2 \tag{6.1}$$

where t_s is the reflection time, h the ice thickness, and c_i the average velocity. Accordingly, a plot of t_s^2 vs. x^2 theoretically yields a straight line, the slope of which is equal to $1/\bar{c}_i^2$.

If only one antenna is moved along the surface, corrections must be made for the slope of the base of the ice relative to its surface. If the angle between the two interfaces is ϕ, then (6.1) becomes

$$\bar{c}_i^2 t_s^2 = (x \pm 2h \sin \phi)^2 + 4h^2, \tag{6.2}$$

from which it follows that

$$\bar{c}_i^2 = \frac{d(x^2)}{d(t_s^2)} \left(1 \pm \frac{2h}{x} \sin \phi\right), \tag{6.3}$$

where the $+$ sign refers to an increasing ice thickness in the direction of the moving antenna. If the slope is small enough so that we can neglect second-order terms, then the error in velocity made by ignoring the slope is given by

$$\frac{\Delta \bar{c}_i}{\bar{c}_i} \simeq \pm \frac{h}{x_{max}} \sin \phi.$$

This shows that the error in the measured velocity increases as the length of the profile, i.e., maximum separation between transmitter and receiver, decreases. This is to be expected, and is an example of the more general fact that velocity measurements from oblique reflection soundings become more and more susceptible to error as the maximum spacing decreases. In a well-designed experiment, the reflections will be carried out to as large a distance as possible – typically the separation can reach about twice the ice thickness before energy diminishes rapidly due to the approach to the limiting ray path, i.e., to a 90° angle of incidence at the surface. We can then approximate the error by letting $x_{max} = 2h$, then $\Delta \bar{c}/\bar{c}_i$ simply equals $\pm \sin \phi$. For example, a bottom slope of 1° will lead to a velocity error of about 1.5 m μs^{-1}. Since the precision of the oblique reflection velocity determinations is about 1 m μs^{-1} under good conditions, even so small a slope could produce a significant error.

For this reason it is important either to conduct a vertical reflection sounding along the line to determine the bottom slope (and, by implication, the bottom topography), or to use an alternative procedure that diminishes the effect of the bottom slope. Such a procedure, called a common reflecting point (CRP) method, involves the separation of both transmitter and receiver equally from a central position, thus maintaining nearly the same reflecting point at the bottom of the ice as the separation increases. In that case, the error in slope determination is reduced to a second-order effect, i.e., the proportional error in velocity depends on $\sin^2 \phi$ rather than on $\sin \phi$, so that a slope of 10° would produce an error in velocity of less than 1%. The CRP method is clearly preferable in principle, but it is more cumbersome in practice and is not always possible to use.

If arrivals at distances larger than the critical distance (the distance at which the lateral wave appears – see Section 6.3) are used, great care must be exercised to avoid contamination by the high-speed lateral wave. The potential effect is particularly strong because of the heavy weight that points at a great distance carry in a least-squares regression line fit to a t^2 vs. x^2 plot. The best procedure is simply to use only arrivals at distances less than the critical one.

The direct wave through the air from the transmitter to the receiver is used to trigger the oscilloscope sweep and, as a reference for all time measurements, the velocity of propagation of this wave must be known to avoid error. Some investigators have stated that antennas must be elevated above the surface (by approximately one-quarter wavelength) to insure that the direct wave travels at the velocity of propagation in air, despite the introduction of uncertainties into the geometry that can also affect the velocity determination (Jiracek, 1967 [110]; Robin et al., 1969 [153]). Such an elevation is, in fact, not necessary. For an infinitesimal dipole lying on the surface of a homogeneous half-space, the expression (first order approximation) for the horizontal component of the electric field, E_y, is

$$E_y = \frac{i\mu_0 \omega I\, dl}{2\pi(k_2^2 - k_1^2)x^2}(ik_2 \exp(ik_2 x) - ik_1 \exp(ik_1 x)),$$

where μ_0 is the permeability of free space, ω is the angular frequency, I is the current in the infinitesimal dipole of length dl, and k_1 and k_2 are the wavenumbers in free space and the lower medium, respectively (Banos, 1966) [30]. The expression shows that there are two separate waves, one travelling in the medium (air) with velocity ω/k_1, and a second wave travelling in the lower medium (ice). On solid ice, the two waves produce an interference pattern as discussed by Hermance (1970) [104] and Annan (1973) [28], but on a snow surface where the density increases, and the velocity thus decreases with depth, the second wave disappears due to downward refraction of the wave. Only one wave is left, travelling along the surface with velocity c. Measurements on Devon Island by Walford et al. (1977) [184] and in Greenland by K.C. Jezek and J.W. Clough (personal communication, 1977), comparing a signal transmitted along a cable with known delay against the same wave in air, confirmed that the air wave near the surface propagates with velocity c (see also discussion of the lateral wave, Section 6.3).

Another possible source of error to be considered is ray-path curvature through the velocity gradient in the firn zone. The magnitude of deviation from straight ray geometry can be estimated by ray-tracing calculations. These show that, for a typical profile on an ice shelf 500 m thick, the error in velocity obtained from a t^2 vs. x^2 plot will be less than 0.5%. It is important to note, however, that the error referred to here is only that from the curvature of the ray path, and does not apply to the effect of the higher velocities near the surface on the average velocity through the ice shelf, an effect which is, of course, still present even when straight ray paths are assumed.

Although the effect of ray-path curvature is small, it cannot be considered completely negligible for the most precise work. Its effect can be removed, however, by a process of 'stripping'. To do this a velocity depth function is calculated, usually from the density depth curve and the equation relating the refractive index and the density given in Section 3.5; ray-tracing techniques are then used to calculate the travel time and range corresponding to the portion of the path in the upper region of varying density. These partial travel times and ranges are subtracted from the total t and x, giving 'stripped' values appropriate to the deeper ice of constant density.

REFLECTIONS FROM THE BASE OF THE ICE

The first oblique reflection experiments were carried out in Greenland in 1963 and 1964 (Robin et al., 1969) [153]. The velocities determined ranged from 153 to 172 m μs^{-1}. The maximum separation was 700 m; the ice thickness was not given.

Many more measurements were carried out in Antarctica in 1964–65 (Jiracek and Bentley, 1971) [111]. Measurements were made in four different regions: the McMurdo ice shelf close to Ross Island, Skelton Inlet (the mouth of Skelton Glacier), where solid ice extends essentially to the surface, on Roosevelt Island ice dome, and on the Ross ice shelf east of Roosevelt Island. Advantage was taken not only of the first reflection, but also of several multiples, i.e., waves that have reflected more than once from the base of the ice with an intervening reflection at the air/firn boundary. As many as four multiple reflections were measured where

SCIENTIFIC RESULTS

TABLE IX

Summary of average wave speeds measured by oblique reflection sounding in polar ice bodies

Station	Ice thickness (m)	Average temperature (estimated) (°C)	Frequency (MHz)	Measured average wave speed (m μs^{-1})	Average wave speed after density correction	Temperature correction to ε'	ε'	Reference
Greenland								
Tuto East	310	−15	35	152–172	152–172	—	3.04 – 3.89	[1]
Camp Century	1390	−20	35	172.1 ± 0.6	171.1 ± 1.0	+0.01	3.08 ± 0.04	[6]
Other Arctic Stations								
Barnes Ice Cap	270		35	171.4 ± 0.8	171.4 ± 0.8		3.06 ± 0.03	[3]
Antarctica, grounded ice								
Roosevelt Island	820	−15	30	174.8 ± 1.0	173.3 ± 1.3	—	3.00 ± 0.05	[2]
Queen Maud Land,								
Station 840	1550	−40	35	172.8 ± 0.7	171.8 ± 1.0	+0.02	3.05 ± 0.04	[3]
Station 840	1550	−40		171.1 ± 1.2	170.1 ± 1.5	+0.02	3.11 ± 0.05	[3]
Dome C	3400	−40	50	170.9 ± 1.7	170.2 ± 2.0	+0.02	3.11 ± 0.07	[6]
Antarctica, floating ice								
McMurdo ice shelf								
Station 203	160	−15	30	177.6 ± 2.0	173.1 ± 2.3	—	3.00 ± 0.08	[2]
Station 204	190	−15	30	178.3 ± 1.2	174.7 ± 1.5	—	2.95 ± 0.05	[2]
Skelton Inlet	770	−15	30	168.5 ± 1.0	168.5 ± 1.0	—	3.17 ± 0.04	[2]
Ross ice shelf (nearby)								
Roosevelt Island	510	−15	30	174.9 ± 0.5	173.0 ± 0.8	—	3.01 ± 0.03	[2]

TABLE IX (Cont'd)

Station	Ice thickness (m)	Average temperature (estimated) (°C)	Frequency (MHz)	Measured average wave speed (m μs^{-1})	Average wave speed after density correction	Temperature correction to ε'	ε'	Reference
Ross ice shelf, RIGGS								
Station H13	770	−15	35	172.2 ± 0.3	170.7 ± 0.4	—	3.09 ± 0.02	[5]
Station H11	630	−15	35	173.3 ± 0.6	171.5 ± 0.8	—	3.06 ± 0.03	[5]
R.I.	620	−15	35	172.9 ± 0.3	170.9 ± 0.5	—	3.08 ± 0.02	[5]
Station N19	570	−15	35	175.3 ± 0.5	173.8 ± 0.6	—	2.98 ± 0.02	[5]
Base Camp	480	−15	35	175.7 ± 0.4	173.5 ± 0.4	—	2.99 ± 0.01	[5]
Station J9	420	−15	50	175.6 ± 2.1	173.6 ± 2.3	—	2.99 ± 0.08	[5]
Station M10	400	−15	35	176.6 ± 0.9	175.0 ± 1.0	—	2.94 ± 0.03	[5]
Station C-16	400	−15	35	176.6 ± 0.9	175.0 ± 1.0	—	2.94 ± 0.03	[5]
1	390	−15	50	177.7 ± 0.3	175.9 ± 0.5	—	2.91 ± 0.02	[5]
2	370	−15	50	178.5 ± 0.2	177.0 ± 0.2	—	2.87 ± 0.01	[5]
Station M14	350	−15	35	178.1 ± 0.7	176.2 ± 1.0	—	2.90 ± 0.03	[5]
Station Q13								
1	340	−15	35	177.2 ± 0.4	175.0 ± 0.4	—	2.94 ± 0.01	[5]
2	330	−15	35	177.9 ± 0.5	176.2 ± 0.7	—	2.90 ± 0.02	[5]
3	330	−15	150	176.4 ± 1.7	173.9 ± 2.2	—	2.98 ± 0.08	[5]
Fimbul ice shelf	310	−15	35	176.0 ± 1.0	172.4 ± 1.0	—	3.02 ± 0.04	[4]

References: [1] Robin *et al.*, 1969 [153]; [2] Jiracek and Bentley, 1971 [111]; [3] Clough and Bentley, 1970 [50]; [4] Van Autenboer and Decleir, 1969 [173]; Decleir, personal communication, 1977; [5] Jezek *et al.*, 178 [109]; [6] Jezek and Shabtaie, personal communication, 1981.

the ice was thin. These multiples were used to minimize the error due to transmission paths in the air, since both antennas were elevated above the snow surface.

The velocity obtained for solid ice was $168.5 \pm 1.0\,\text{m}\,\mu\text{s}^{-1}$ on Skelton Inlet. Since there is no significant firn zone in that region, no correction was needed for velocity variations near the surface. In the other areas, measured velocities ranged from 174.8 to $178.3\,\text{m}\,\mu\text{s}^{-1}$. These values have been corrected for near-surface densities by Jezek et al. (1978) [109], yielding values between 173.0 and $174.7\,\text{m}\,\mu\text{s}^{-1}$, as shown in Table IX.

During the 14th SAE (1969) wave speeds were measured at 7 sites (strain polygons) in the vicinity of Molodezhnaya Station by means of a RV-10 radar and a laser (see Section 4.3). Results along one line are given in Table X. Results of $t_s^2 = f(x^2)$ measurements along all 7 lines are shown in Figure 39, along with straight lines fit by least-squares. The speeds obtained in the coastal zone were 167.8, 168.6, 170.0, and $178.0\,\text{m}\,\mu\text{s}^{-1}$, and $184.5\,\text{m}\,\mu\text{s}^{-1}$ was measured 45 km from the coast.

TABLE X
Parameters determining wave velocity

x,	m:	0	50	100	150	200	250	300	350
t_3,	μs:	2.25	2.0	1.85	1.72	1.7	1.7	1.75	1.83
t_x,	μs:	0	0.165	0.33	0.5	0.66	0.83	1.0	1.165
t_s,	μs:	2.25	2.165	2.18	2.22	2.36	2.53	2.75	2.995
$x^2 \times 10^{-4}$, m^2:		0	0.025	1.0	2.25	4	6.25	9	12.25
t_s^2,	μs^2:	5.08	4.7	4.75	4.93	5.41	6.4	7.57	8.95

In the 15th SAE (1970) wave speed measurements were continued in the Molodezhnaya Station area. Three polygons with ice thicknesses from 300 to 500 m were

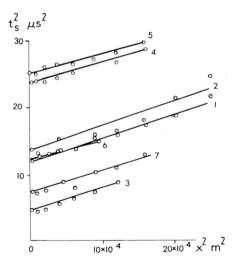

Fig. 39. Velocity measurement by t_s^2 vs. x^2 plot. $\bar{c}_i$: lines 1 and 2 – $168.6\,\text{m}\,\mu\text{s}^{-1}$; line 3 – $178.0\,\text{m}\,\mu\text{s}^{-1}$; lines 4 and 5 – $184.5\,\text{m}\,\mu\text{s}^{-1}$; line 6 – $172.0\,\text{m}\,\mu\text{s}^{-1}$; line 7 – $167.0\,\text{m}\,\mu\text{s}^{-1}$.

chosen, and laser measurements were made along three lines at each polygon. Echoes were received at dipole antennas oriented parallel and perpendicular to the transmitting dipole axes. The results are listed in Table XI.

TABLE XI

Number of the measurement line	1.1	1.2	1.3	2.1	2.2	2.3	3.1	3.2	3.3
Mean ice thickness along the route (m)	300	300	309	426	568	465	367	384	375
Wave velocity in ice (m μs^{-1})	163	170	156	185	170	183	179	172	172

In the 16th SAE (1971) wave speed measurements were carried out by means of a 60–67 radar. Since it is a powerful instrument, measurements were carried out in areas with comparatively greater ice thicknesses. The results obtained at six polygons south of Molodezhnaya Station are shown in Table XII. At each polygon except the second one, measurements were carried out along two lines, one to the north and one to south from a chosen point.

TABLE XII
Wave velocity in thick ice (Molodezhnaya Station area)

	Number of the polygon					
	1	2	3	4	5	6
Mean ice thickness along the profile (m)	820.0	597.0	860.0	650.0	133.0	313.0
Wave velocity in ice (m μs^{-1})	172.0	—	182.0	162.5	153.0	160.0
	160.5	167.0	172.5	169.0	154.0	170.0

In the 21th SAE and 22nd SAE (1976–77) oblique reflection measurements were performed. They were made on an ice dome 50 km south of Novolazarevskaya Station, and in the area of Druzhnaya Base on the Filchner ice shelf; the ice thickness on the ice dome near Novolazarevskaya averaged 1000 m, whereas on the Filchner ice shelf it was measured to be about 300 m. At the ice dome 6 measurements of $\bar{c}_i$ were made along 4 lines, each 500 m long. Four measurements also were carried out on the Filchner ice shelf. The results indicate wave speeds of 160 m μs^{-1} through the dome ice and 172 m μs^{-1} through the shelf ice. These values are comparable to those listed in Table XI and XII; they differ from those in Table VIII.

The oblique reflection technique has been used on Vavilov Glacier, Severnaya Zemlya, since 1975. The 100-75 radar (see Section 4.4) was used for these experiments. Fifteen series of measurements were completed. The results of seven measurements of the first series yield a graph in t_s^2 vs. x^2 coordinates that is not a straight line but a curve with a well marked minimum (Figure 40). This shows the chosen model to be inadequate and (6.1) to be incorrect. It follows from the graph that the downward portion of the curve corresponds to a false speed, a horizontal line corresponds to an infinitely large speed, and only the upward part of the curve

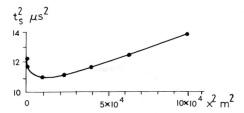

Fig. 40. Curve showing the $\bar{c}_i$ anomaly in oblique-reflection sounding results.

seems to correspond to a real speed. There is no explanation yet, though similar features were found during the studies on Antarctic glaciers as well.

Three different ways to process the results were proposed by Fedorov (1978) [23]. His comparative study (Table XIII) indicates a large variation in estimated values of $\bar{c}_i$. Many of the quantities obtained are lower than the velocity in pure ice (density $0.92\,\mathrm{mg\,m^{-3}}$), $167\,\mathrm{m\,\mu s^{-1}}$. For the oblique soundings of series 14 at critical distances, the value of $\bar{c}_i$ obtained is close to the speed in pure snow. For measurement site II, where the largest number of measurements were made, the range of most probable speeds is $\bar{c}_i = 170–175\,\mathrm{m\,\mu s^{-1}}$. Wider ranges of values of $170–180\,\mathrm{m\,\mu s^{-1}}$ and $170–190\,\mathrm{m\,\mu s^{-1}}$ were estimated for sites I and III. The

TABLE XIII

Mean wave velocities through Vavilov Glacier ($\mathrm{m\,\mu s^{-1}}$)

Measurement site	Series	Ice thickness (m)	Oblique sounding with a time delay			Oblique sounding at critical distances
			Graphical averaging	Calculated by distant points	Calculated by characteristic points	
I	1	290	169.0	153.5	164.5	
	2	295	196.0	—	194.0	
	3	277	—	—	—	
II	4	487	—	—	—	157.0
	5	487	197.0	—	179.0	183.0
	6	487	—	—	—	109.0
						144.0
						150.0
	7	487	—	—	—	158.0
						146.0
	8	487	—	—	—	171.5
	9	487	199.0	—	180.0	—
	10	487	—	—	—	175.0
	11	487	183.0	—	173.0	—
	12	487	—	—	—	188.0
	13	487	192.0	—	173.0	200.0
III	14	315	—	—	—	230.0
	15	315	147.0	155.0	153.0	—

deviations of the measured $\bar{c}_i$ values towards larger quantities can be attributed to the effects of a snow–firn layer, the thickness of which was not measured. Nevertheless, the number of measurements is quite sufficient to assume $\bar{c}_i$ equal to 168 m μs^{-1}.

Velocity determinations on the much thicker ice in the interior of Queen Maud Land on the East Antarctic ice sheet were carried out during the ground traverses of 1965–66 and 1967–68 (Clough and Bentley, 1970) [50]. At 79°S, 07°W, two wide-angle profiles, one with a common reflecting point and one with a fixed end, were completed along the same line. The values obtained for the velocity were 172.8 ± 0.7 m μs^{-1} for the CRP profile, and 171.1 ± 1.2 m μs^{-1} for the fixed end profile, the latter having been corrected for the effect of the subglacial topography. Correction for the near-surface yields 171.8 ± 1.0 m μs^{-1} and 170.1 ± 1.5 m μs^{-1}, respectively, for the solid glacier ice.

A CRP profile was also completed on the Barnes Ice Cap on Baffin Island during May 1967 (Clough and Bentley, 1970 [50]). The mean ice cap temperature was about −12°C. The velocity, 171.4 ± 0.8 m μs^{-1}, which requires no correction for near-surface variations, is essentially the same as that for Queen Maud Land. The overall average of 171 ± 2 m μs^{-1} for Queen Maud Land and the Barnes Ice Cap is equivalent, after a small temperature correction, to $\varepsilon' = 3.12 \pm 0.05$ at a temperature of −10°C.

More recently, CRP profiles have been completed on the inland ice at Camp Century, Greenland, and Dome C, East Antarctica (K.C. Jezek and S. Shabtaie, personal communication, 1981). At both sites bottom topography was determined by profiling; corrections were found not to be necessary.

The Camp Century experiment involved measurements at over 60 distances to a maximum exceeding 6 km, providing a very well-determined line (Figure 41). In

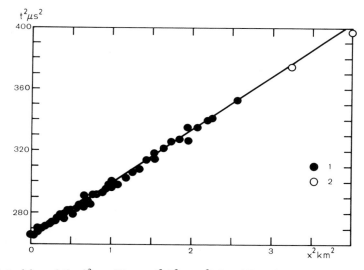

Fig. 41. Plot of (travel time)^2 vs. (distance)^2 (t^2 vs. x^2) for oblique ice-bottom reflections at Camp Century, Greenland. Open circles were not used in the least-squares fit to the slope. (K.C. Jezek and S. Shabtaie, personal communication, 1981.)

calculating the velocity, the two most distant points were neglected, as they fall below the average least-squares fit, and may be contaminated by the lateral wave, since the ice thickness is only 1390 m. The fit to the x^2 vs. t^2 plot yields a wave speed of 171.9 ± 0.7 m μs^{-1}; corrected for the firn layers that becomes 171.1 ± 1.0 m μs^{-1}.

At Dome C measurements reflections were recorded at over 20 distances out to 7.6 km on ice 3400 m thick. The lateral wave is not a factor to be considered. The uncorrected wave speed was 170.9 ± 1.7 m μs^{-1}; corrected, that becomes 170.2 ± 2.0 m μs^{-1} in solid glacier ice.

The close agreement between measured wave speeds in glacial ice from the four CRP profiles on grounded ice is striking – all are within the range 170–172 m μs^{-1}, and the average is 171.1 ± 0.7 m μs^{-1}. The corresponding value of ε', after a small temperature correction to $-15°C$, is 3.083 ± 0.025. That is significantly lower than most laboratory measurements, but not significantly different from the measurements on deep ice from the Byrd Station core hole (see Section 6.1).

Surprisingly high velocities were revealed also by extensive measurements on the Ross ice shelf (Jezek et al., 1978) [109]. The results of the 13 wide-angle velocity profiles on the Ross ice shelf are summarized in Table IX. Average velocities through the entire ice shelf and velocities found after 'stripping', differing from the uncorrected values by about 2 m μs^{-1}, are both given, along with values of ε' calculated from the corrected velocities. No corrections for temperature or frequency have been applied. Mean annual surface temperatures vary from place to place only over a 5° range (Thomas, 1976 [170]; Crary et al., 1962 [60]); average temperatures in the ice shelf will differ by only about half that. Since $\partial \varepsilon'/\partial T$ is less than 10^{-3} deg^{-1} (section 3.2), temperature corrections are negligible.

The data listed in Table IX include error estimates calculated from linear regression fits to the slope and intercept on the t^2 vs. x^2 plots. They reflect a high precision in the data. The reproducibility is nearly as high, as can be seen by comparing the two values at C-16, and the three at Q13. A reasonable estimate of the standard error in measured velocity is ± 1 m μs^{-1} (except at J9, where the scatter of the data points was wide). This corresponds to about ± 0.03 for the values of ε' and about ± 5 m for the thickness.

Also included in Table IX is the result of another field measurement, this one on the Fimbul ice shelf on the Atlantic side of the Antarctic, where again the measured velocity was high (Van Autenboer and Decleir, 1969 [173]; H. Decleir, personal communication, 1977). Jezek et al. (1978) [109] applied density corrections to Decleir's value to obtain a value of 172.4 ± 1.0 m μs^{-1} ($\varepsilon' = 3.02 \pm 0.04$).

The large variation in values of ε' in the ice shelves shows clearly in Table IX. The range is from 2.89 (average at C-16) to 3.09 (H13). At present, the cause of the differences is not known. One possibility is some inherent error in the wide-angle technique that has not been recognized. One hypothetical model is that the reflection surface gradually migrates upward as the angle of incidence increases. As much as 15 to 20 m of migration would be needed – perhaps that could somehow occur if there is a zone of salty ice frozen on to the base of the shelf.

Some support for that idea comes from a plot of the geographical distribution of ε' (Figure 42). Values decrease as the distance downstream from the grounding line becomes greater, consistent with a model of bottom freezing, and then

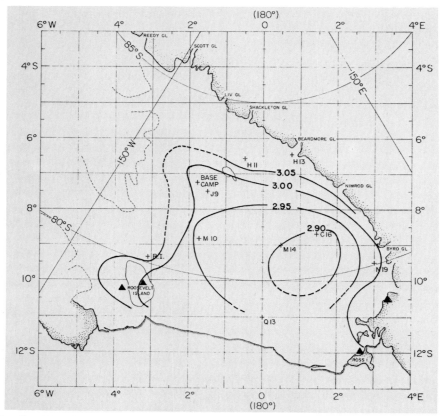

Fig. 42. Map of ε' in the Ross ice shelf, as measured from oblique reflections and corrected for near-surface densities. +: RIGGS stations, ▲ from Jiracek and Bentley (1971). Grid coordinates are centered on the South Pole, parallel and perpendicular to the Greenwich Meridian, with grid north toward Greenwich (from Jezek et al., 1978).

(perhaps) increase again toward the ice front in a region where bottom melting is likely to occur.

However, there is very little evidence from the actual oscilloscope photographs to support such a migration, nor is it easy to see how any migration could occur gradually rather than in a very few finite jumps. Furthermore, the reflection coefficient at any boundary should vary only gradually out to angles of incidence greater than included in these experiments. Thus this model is difficult to support quantatively.

If the variations in the measured values of ε' do not arise from some systematic error in the wide-angle reflection technique, then they must represent real variations in the permittivity of the solid shelf ice.

Unfortunately, no satisfactory explanation of such low values of ε' has yet been put forward. If there are impurities in the ice, ε' should increase rather than decrease. Nor is there reason to believe that the explanation lies in crystal anisotropy, since any difference in ε' parallel and perpendicular to the crystallographic

c-axis must, from laboratory measurements, be less than 0.5% (see Section 3.2).

Whatever the explanation for the variations in the measured values of ε' they appear to arise not simply from ordinary experimental errors, but from some real physical characteristics of glacial ice. Experiments on deep ice cores from the ice shelves and inland ice sheets are needed to determine whether those characteristics relate to propagation paths or to real permittivity differences in cold glacier ice. (See Note on p. 120.)

REFLECTIONS FROM INTERNAL LAYERS

The velocity variation in the upper part of an ice sheet is of interest because of its dependence on density and on geometry of the ice–air mixture. One can, in principle, take advantage of the succession of internal layers which have been found to exist down to a depth of 1000 m or more in the ice and employ the oblique reflection method. At the same time, the attitude and horizontal continuity of the reflecting surface can be measured.

Initial measurements of this kind were carried out during the 1965–66 and 1967–68 ground traverses in Queen Maud Land (Clough and Bentley, 1970) [50]. Unfortunately, the first two microseconds of the return signal are obscured by the direct pulse through the air. Consequently, any reflections within the upper 200 m

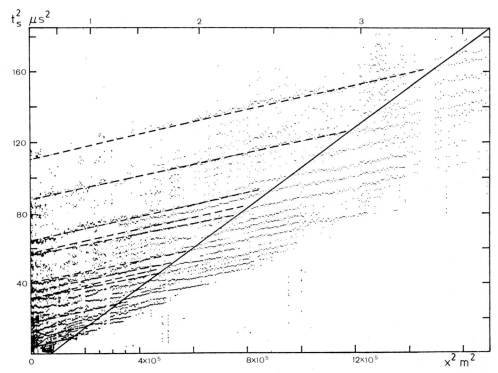

Fig. 43. Plot of t^2 vs. x^2 for oblique reflections from internal layers at Byrd Station, Antarctica. Dashed lines were used to calculate the average wave speeds in Figure 44. (From Clough, 1974.)

of the ice sheet are masked and results cannot be obtained for this zone. For the rest of the records, some of the events are easy to correlate from one record to the next whereas others are more variable in nature and may be present only in a portion of the records. To analyze the data, all distinguishable reflections were included on a t^2 vs. x^2 plot. Connecting lines were drawn where the corresponding correlation could be made on the oscillograms; thus, some of the lines extend across the whole range of observations whereas other are only partial. Mean velocities were calculated from the best determined lines and plotted as a function of depth. Error bars were determined graphically by estimating the maximum and minimum slope that could reasonably be assigned to a given set of points.

The resolution of the experiment was insufficient to detect real deviations in velocity from a mean curve based on an estimated density–depth function. There was good enough agreement, however, to prove the reality of some 15 internal reflectors between depths of 250 and 1250 m, and to demonstrate the horizontal continuity of some of the reflectors over distances of hundreds of meters.

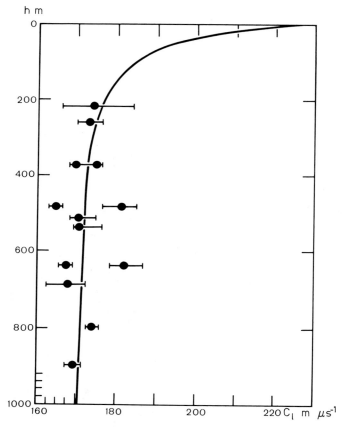

Fig. 44. Average wave speed from the surface to various internal layers in the ice sheet at Byrd Station, Antarctica, as a function of the depth to the layer. The solid line is a calculated curve of average velocity vs. depth from measured densities. (From Clough, 1974.)

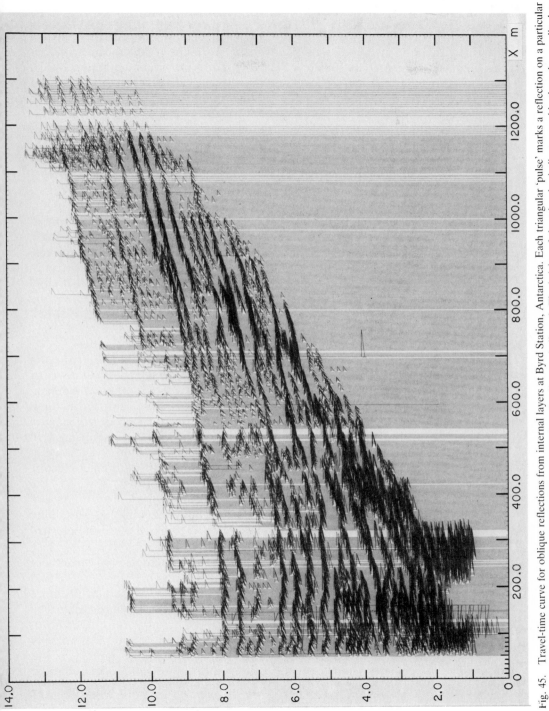

Fig. 45. Travel-time curve for oblique reflections from internal layers at Byrd Station, Antarctica. Each triangular 'pulse' marks a reflection on a particular record; each vertical line corresponds to a separate record at the distance indicated. The heights of the triangles indicate roughly the echo amplitudes. (From Clough, 1974.)

In the hope of providing enough resolution to predict density–depth curves from a measured velocity vs. depth function, a closely detailed survey was carried out in 1970 near Byrd Station (Clough, 1974) [52]. CRP profiles were made to distances of 1400 m at 2 m intervals. An x^2 vs. t^2 plot is shown in Figure 43, and the mean velocities obtained therefrom are shown in Figure 44. Error bars were determined in the same way as for the Queen Maud Land data. Even with this extremely detailed work, the scatter in the celocities has not been improved over the Queen Maud Land work.

The reason for this can be found by examining the travel time plot for one of the Byrd Station profiles (Figure 45). Several reflection curves can be seen on the travel time plot but there is considerable ambiguity when trying to follow a particular curve through the plot. It is the discontinuity of reflections and the irregularity and scatter of the points evident on the plot that prevent better velocity determination. Although these discontinuities and irregularities may be attributed in part to instrumental problems, the data certainly do not support an interpretation for the internal layering that would require the layers to be smooth, horizontal, and continuous over large distances. Detailed reflection profiles on the internal layers collected by K.C. Jezek and C.R. Bentley in 1978–79 and S. Shabtaie in 1978–80 at Dome C on the East Antarctic ice sheet provide further confirmation of this.

COMPARISONS WITH SEISMIC SOUNDING

The electromagnetic wave velocity could, of course, be determined wherever radar soundings have been carried out at a location where the ice thickness is known by some other means. In principle, seismic soundings provide the required independent determination of ice thickness. Since there are many more seismic soundings on ice sheets than there are drill holes through the ice, such comparisons are relatively numerous. However, it has been clear from the earliest comparisons that discrepancies between the seismic and radar reflection results often arise. Some of these discrepancies have been attributed to the possibility of different reflecting surfaces for the seismic and electromagnetic waves. More often, however, the differences have been attributed to variations in wave speed from place to place in the ice. There is general agreement among glacio-geophysicists that the variability is primarily attributable to the seismic wave velocities rather than to the electromagnetic wave velocities.

The first direct comparisons between the two different methods was carried out as part of an international experiment in glacier sounding in Greenland in 1963 and 1964 (Waite, 1966) [180]. Maximum discrepancies amounted to less than 3% in ice between 200 and 1400 m thick. The mean difference in depth without regard to sign was ±15 m. Taking account of sign, the mean was +9 ±3 m (Evans, 1967 [76]), indicating that the seismic velocity used was ~1% too low, or the radar velocity 1% too high, or some combination of the two. Nevertheless, the agreement was impressive. A similarly good agreement was found between seismic and radar measurements on the Penney ice cap on Baffin Island (Weber and Andrieux, 1970 [187]).

In the 11th SAE (1966) B.A. Fedorov had made 24 radar measurements along a 100 km route southward from Mirny observatory, along the meridian. Based on these measurements, the glacier profile and the bedrock relief were drawn. In the 2nd SAE (1957) seismic sounding of the glacier was done on the same route. Seismic data were also used to draw the profile and the bedrock relief. The two graphs are almost identical. Both of them show bedrock sections below sea level for 50–60 km south of Mirny. Farther south, 100 km from Mirny, the bedrock rises and becomes higher than sea level. However, details of the relief do not coincide and the glacier thickness measured by seismic sounding almost everywhere is less than that measured by radar. The exceptions are near 30, 70, and 100 km. At these points the glacier thickness measured by the seismic technique is more than that measured by radar.

In the 18th SAE (1973) on the same route (Mirny to 153 km south), combined radar and seismic measurements of glacier thickness were carried out. Measurements were made at the same points. The results are given in Table XIV.

It is seen from table XIV that the glacier thickness measured by the seismic method is considerably larger than the thickness measured by radar (operating frequency 213 MHz). At points 2, 4, 45, and 47, the thickness difference exceeds a hundred meters. Such discordance is very difficult to explain. Perhaps the error arises due to an incorrect choice of seismic wave velocity. The electromagnetic wave velocity was not measured in this experiment, but was taken to be 167 m μs^{-1}

TABLE XIV
Glacier thicknesses from seismic and radar measurements

	Point number						
	1	2	3	4	45	46	47
Distance from Mirny (km)	13	27	50	55	80	102	125
H_r (m) (radar)	545	872	824	735	1199	1254	1798
H_s (m) (seismic)	615	1087	872	813	1386	1330	1931

In the 1964–65 Antarctic field season, velocity comparisons were made at South Pole, on Roosevelt Island, and on the Ross ice shelf (Jiracek, 1967 [110]; Jiracek and Bentley, 1971 [111]). At South Pole Station, where the ice thickness calculated from seismic measurements, using a wave velocity of 3850 m s^{-1}, is 2850 ±30 m, the echo time was 33.0 μs. Drewry (1975) [69], however, reports an echo time recorded in 1971 of 34.5 μs, less than a kilometer away. Since both seismic and radar echo times vary an irregular subglacial topography is indicated, making detailed comparisons impossible.

On Roosevelt Island many direct comparisons were made, and the agreement was not good. Seismic depths (Hochstein, personal communication) were 50 to 100 m less at all stations than those determined by the radar technique. Furthermore, average velocities for both seismic waves (3800 m s^{-1} in the ice) and radar waves (173 m μs^{-1}) were calculated by oblique reflection sounding at the same location. In this case then the consistent difference between the results casts uncertainty on the identity of the reflecting surface, suggesting that the 'floors', as determined by seismic and radio reflecitons, may not coincide. For further discussion of this point see Section 6.10.

Ice thickness arrived at by radar and seismic sounding on the Ross ice shelf east and west of Roosevelt Island were in satisfactory agreement.

Much more extensive measurements were carried out in Antarctica during the 1965–66 and 1967–68 ground traverses in Queen Maud Land (Clough and Bentley, 1970) [50]. The travel times for vertical electromagnetic and seismic reflections at 13 stations, covering ice thicknesses between 2200 m and 3400 m, show a reasonably consistent relationship (Figure 46). Travel times for both types of reflections have been corrected for the effect of near-surface velocity structure. The least-square regression line equation fit to the data is

$$t_e = (22.8 \pm 1.0) t_p + 2.0 \pm 1.5, \qquad (6.4)$$

where t_e and t_p are the electromagnetic and compressional wave travel times in microseconds and seconds, respectively. For a normal seismic velocity $c_p = 3900 \text{ m s}^{-1}$, (6.4) yields an electromagnetic wave velocity of $170 \pm 7 \text{ m } \mu\text{s}^{-1}$ ($\varepsilon' = 3.11 \pm 0.26$). This agrees with other determinations of electromagnetic wave velocity and dielectric constant, but clearly the precision is not sufficient to gain new velocity information. Furthermore, a problem remains concerning the positive t_e intercept in (6.4) that may affect the velocity ratio.

For a simple model of the ice sheet as a homogeneous, isotropic slab of ice lying on a discrete reflecting base, t_e should clearly be directly proportional to t_p. The t_e intercept of $+2 \mu\text{s}$ seen in Figure 46, however, appears to be real. Similar time differences between the two techniques have reported by Jiracek (1967) [110], Robin et al. (1969) [153], and D. Carter (personal communication). Several

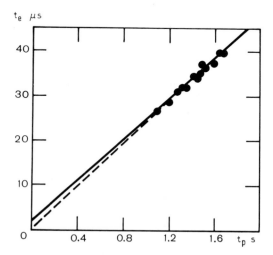

Fig. 46. Radar reflection time (μs) vs. seismic reflection time (sec) in Queen Maud Land, Antarctica. Solid and dashed lines are explained in the text. (From Clough and Bentley, 1970.)

possible explanations should be considered: (1) Systematic errors in time measurements are responsible. This may be true in part, but it is unlikely that these errors could be as large as 2 µs. The error in seismic travel time may be taken as about 0.005 s, which corresponds approximately to a 0.1 µs change in intercept. The resolution of the electromagnetic travel times is estimated to be 0.5 µs or better. (It is assumed that the time required for the signal to pass through the receiver and display equipment is the same for the echo and the initial pulse.) It is difficult, therefore, to see how systematic errors could cause more than a quarter of the observed intercept. (2) There is a real difference in reflecting boundary for the two wave types, the electromagnetic reflector lying deeper. This possibility is discussed by Jiracek (1967) [110] in the case of Roosevelt Island, where there is an additional complication in an apparent requirement for the wave velocity to increase beneath the ice. He found a satisfactory model difficult to devise. Even without the velocity restriction in the present case, it is still necessary to postulate a layer more than 100 m thick, which must present a low electromagnetic, but high acoustic, impedance contrast with the ice. (Geometrical properties of the boundaries cannot be called upon because of the near equality of seismic and electromagnetic wave lengths – about 15 m and 5 m, respectively.) Suitable earth material does exist, e.g., high-aluminum-clay-permafrost (Cook, 1960) [58], but there is no geological support for the existence of such a thick, extensive, and uniform layer of specialized composition. The hypothesis of different, widely separated reflecting boundaries does not appear tenable for Queen Maud Land. (3) Either c_e or c_p, or both, differ considerably from the expected value. Forcing the regression line through the origin yields a slope of 24.3 µs s^{-1}. Then, for $c_e = 170$ m µs^{-1}, c_p would be 4130 m s^{-1}, whereas $c_p = 3900$ m s^{-1} would imply that $c_e = 160$ m µs^{-1} ($\varepsilon' = 3.5$). The only conceivable cause of such discrepant velocities would be anisotropy.

The compressional wave velocity in ice is strongly affected by anisotropy, the maximum value along the c-axis exceeding the polycrystalline average by more than 200 m s^{-1}. Strong anisotropy has been observed in the lower part of the ice column in the deep drill hole at Byrd Station (Gow et al., 1968) [87], and has been shown by seismic measurements to be widespread in West Antarctica (Bentley, 1971) [33]. In the Byrd core a strong preferred alignment of c-axes near the vertical was observed.

Suppose the East Antarctic ice sheet to be made up partly of isotropic, and partly of anisotropic ice. If, in the investigated region of Queen Maud Land, ice thickness changes from place to place reflect changes in the thickness of the isotropic fraction, the regression slope of t_e on t_p would equal c_p/c_e in isotropic ice, and the regression line would have a non-zero intercept, as in Figure 46.

To attempt a quantitative fit to the data, consider a specific model: a layer with a constant thickness of 2500 m in which there is perfect vertical alignment of c-axes ($c_p = 4100$ m s^{-1}), and a layer of isotropic ice of variable thickness ($c_p = 3900$ m s^{-1}). The corresponding t_e vs. t_p relation would then follow the solid line in Figure 46 for $t_p > 1.2$ s, and the dashed line for $t_p < 1.2$ s, the latter segment referring, of course, to a total ice thickness of less than 2500 m, so that the anisotropic ice thickness would perforce be changing. With this model a positive intercept of 0.5 µs still remains – this much, perhaps, could be attributed to systematic errors.

The model described is too extreme to seem very satisfactory. It is worth noticing, however, that, in many instances, seismic velocity variations in West Antarctica can be explained only by extreme models. Furthermore, if the bottom several hundred meters of ice are isotropic, as at Byrd Station, major changes in total ice thickness over rugged subglacial terrain might indeed occur primarily in the isotropic layer. Thus, the model at least has the merit of qualitative support from independent glaciological data. We conclude that seismic anisotropy, together with some systematic errors in time measurement, is the most likely explanation for the observed intercept.

Drewry (1975) [69] has made a detailed comparison between radar soundings and seismic soundings based on airborne radar flights in 1971–72 and seismic soundings from ground traverses between 1958 and 1965. Using a standard velocity of $169 \, \text{m} \, \mu\text{s}^{-1}$ for the radar data, he shows that the average differences were less than 4%. This is a useful result in verifying the ice thickness measurements by the two methods, but the comparison is clearly insufficient to yield useful velocity information.

All comparisons on the grounded ice sheet suffer from uncertainty about the common identity of the reflecting surfaces for the electromagnetic and acoustic waves. That uncertainty is not present for measurements made on a floating ice shelf. Robertson (1975) [146] reports on seismic and radar reflections from the Ross Ice Shelf Geophysical and Glaciological Survey results of 1973–75. The results are displayed in Figure 47, where the ice thickness, h, computed from radar sounding ($c_i = 174 \, \text{m} \, \mu\text{s}^{-1}$) is plotted versus Δh, the difference between radar and seismic ice thickness (the latter calculated using a velocity in ice of $3800 \, \text{m} \, \text{s}^{-1}$). The estimated error in measuring h by either seismic or radar means is $\pm 10 \, \text{m}$, so the estimated error in Δh is $\pm 14 \, \text{m}$. The least-squares regression line forced through the origin is

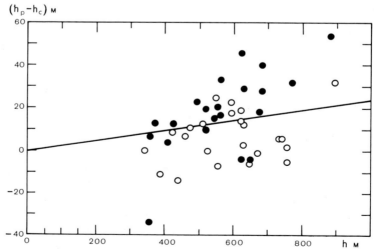

Fig. 47. Difference between ice thickness as determined from radar reflections, h_r, and from seismic reflections, h_s, as a function of ice thickness for the Ross ice shelf. Solid and open circles are from the 1973–75 seasons, respectively. (From Robertson, 1975.)

$$\Delta h(\pm 14) = 0.024(\pm 004)h,$$

indicating that the P-wave velocity is a little more than 2% higher than expected. The results support a preferred vertical orientation of c-axes in the ice shelf, though it must be noted scatter of the data is large and the conclusion no more than a general impression of overall average conditions.

There appears to be a significant difference between the degrees of c-axis orientation found in the 1973–74 survey area and the 1974–75 survey area. The least-squares fit to 20 determinations from 1973–74 is

$$\Delta h(\pm 13) = 0.038(\pm 0.005)h,$$

and to 24 values from 1974–75 is

$$\Delta h(\pm 11) = 0.012(\pm 0.004)h,$$

corresponding to 4% and 1% higher P-wave velocities respectively. In addition, many variations between adjacent individual stations exceed 15 m in magnitude, and apparently are significant. No prominent regional pattern, however, is evident in the distribution of individual values.

COMPARISON OF RESULTS

Figure 48 shows a compilation of all the different field measurements of velocity. Also shown for comparison are measurements taken in the laboratory on samples of polar ice, corrected to $\rho = 0.97$ mg m^{-3} and $t = -15°C$. It is clear that most of the field measurements, taken by different investigators with different equipment, have yielded velocities higher (ε' less) than the laboratory measurements. This is especially striking for the ice shelf data. Particular notice should be taken of Jiracek's determination on the McMurdo ice shelf (the two open circles at ice thickness <200 m), since they employed multiple reflections, thus eliminating uncertainties about the direct wave travel time (Jiracek and Bentley, 1971) [111], and to the remarkable agreement at Camp Century between values obtained by different methods.

Standing in strong contrast to all the oblique reflection measurements, however, is the single determination by the interference method on the Devon Island ice cap (Robin, 1975) [151]. The accuracy of that value appears incontrovertible, suggesting that there is some systematic error in the oblique reflection measurements. The result on the ice at Skelton Inlet ($c_i = 168.5$ m μs^{-1}) where there is no firn zone, and also the decreasing velocity as the ice gets thinner, suggest that improper allowance for the firn layers might be distorting the results. However, a simple calcultion for Camp Century, as an example, shows that for c_i to be 168.5 m μs^{-1}, the mean speed in the upper 200 m would have to be 200 m μs^{-1}, corresponding, according to any accepted relationship between density and wave speed, to an effective mean density of only 0.6 mg m^{-3}, which is patently absurd.

Thus, we are left at present with a contradiction that only more laboratory measurements and field work can resolve. (See note on p. 120.)

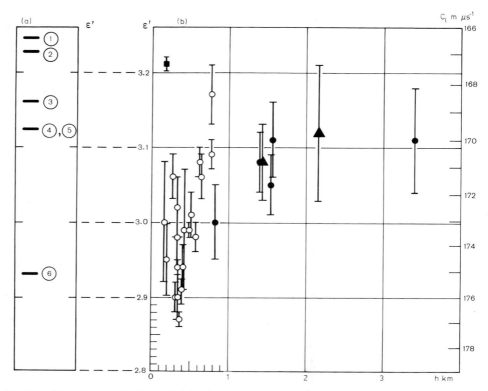

Fig. 48. Plot of determinations of ε' in glacial ice, as calculated from field velocity measurements. Solid circles: oblique reflections on grounded ice; open circles: oblique reflections on floating ice; solid square: interferometry; solid triangles: reflection times at deep drill holes. All measurements have been corrected to T = $-15°$C. Also shown for reference are laboratory measurements carried out on polar ice samples, corrected to $-15°$C, from Westphal (Jiracek, 1967): ①-④; from Fitzgerald and Paren (1975): ⑤; from Paren and Glen (1978): ⑥.

6.3. Lateral Waves

Practically all the discussions thus far have been related to waves that travel through an ice body, reflect from a boundary within, or at the base of the ice, and return to the surface. There are, in addition, several waves that are of interest more for their propagation characteristics than as a means of studying the ice. For some time there were questions as to whether the wave traveling horizontally between two antennas resting on the surface of an ice sheet would propagate at the speed in air or would be slowed down in a way corresponding to the behavior of a surface wave at the boundary between space and a conductor. Clough (1974, Jezek *et al.*, 1978) [52, 109] showed, both theoretically and experimentally, that the former is the case, i.e., that a direct wave does propagate with the speed in the air. As such, it can be used reliably as a measure of initiation time of a pulse when the transmitting and receiving systems are operated separately.

An electromagnetic wave incident from within the ice at the boundary between the ice or firn and the air will theoretically be refracted according to Snell's Law. At the critical angle, this gives rise to the so-called 'lateral wave'. The low-amplitude lateral wave is usually obscured by other waves when continuous wave sources are used. But by using the pulsed source of a radar system and continuously recording echoes reflected from within the ice as a function of incidence, records can be produced which clearly show the lateral wave. Clough (1976) [203] presents a photographic recording of an entire oblique reflection profile on the Ross ice shelf in one picture (Figure 49). The photograph shows the lateral wave tangent to the reflection at distances beyond the critical angle. (For reasons that are not known, the lateral wave arrival is slightly delayed relative to a line tangent to the reflection at the critical angle.) The reflected arrival diminishes rapidly with increasing distance beyond a maximum distance that is determined by the wave velocity in the uppermost firn. The lateral-wave amplitude decreases more slowly, so it becomes equal to (Figure 50), and then greater than, the reflected-wave amplitude. Eventually, the lateral wave becomes the only one recorded.

The radiation pattern from an antenna resting on the firn surface exhibits a 10 dB power amplification at an outgoing angle equal to the critical angle for the ice–air boundary. An effect of this can be seen in Figure 49. There are some very faint reflection curves observed between 2.0 and 5.0 μs that arise from internal reflectors. As the horizontal range is increased, successively deeper reflectors are briefly

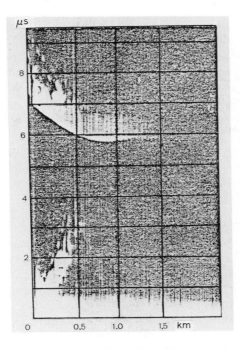

Fig. 49. Reduced travel-time curve of oblique reflection profile, Ross ice shelf, Antarctica. Visible are the direct wave (bright line across the bottom of the figure), the reflected wave, the lateral wave (1.2–1.5 km, just below 6 μs), and the line formed from critical-angle internal reflections (see text). (From Clough, 1976.)

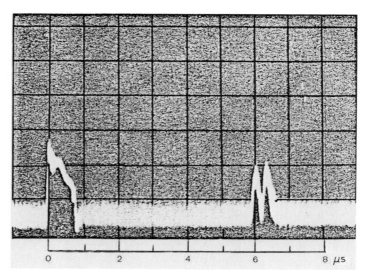

Fig. 50. A-display taken at a distance of about 1.3 km on the profile of Figure 49 showing the direct wave and the lateral and reflected waves at 6.0 and 6.2 μs respectively. (From Clough, 1976.)

illuminated at the angle of maximum power. The result is seen in Figure 49 as a bright zone extending from the origin toward the main reflection curve at a distance of ~1.0 km. The brightening does not show up well in the case of the main reflection curve, probably because of non-linearities in the recording process.

A modification of the reflection experiment was performed by Clough and Bentley (1970) [204] at Byrd Station, Antarctica. Travel-time profiles were made for waves traveling one-way paths from a transmitter on the surface to receivers placed approximately 42 and 21 m below the surface. These profiles should yield velocities through the upper snow layers, and should record the occurrence of electromagnetic wave propagation along the surface at the free-space velocity. The travel-time plot for one of these profiles (Figure 51) shows the direct wave (equivalent to half a wide-angle reflection) at transmitter–receiver spacings of less than 90 m, and the lateral wave at greater distances. The measured speed of the lateral wave was 302 ± 6 m μs^{-1}.

As previously mentioned, a surface wave does not propagate far on the upper surface of an ice sheet, so it cannot be observed along a direct path between transmitting and receiving antennas because it is obscured by the tail of the direct wave. A wave that is probably equivalent, i.e., that travels in the uppermost part of the firn zone, can be produced, however, when the angle of incidence of a reflection from the base of the ice reaches grazing incidence in the uppermost firn layers. In this case, there is no other obscuring energy arriving, and waves which had traveled along the reflection path and then horizontally within the uppermost firn were observed at several sites on the Ross ice shelf (Jezek et al., 1978) [109]. These waves are attenuated rather rapidly, however, and are not seen at large distances beyond the critical angle. Nevertheless, if they are not properly taken into account,

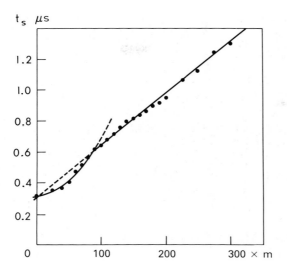

Fig. 51. Travel-time plot for oblique propagation between the surface and a receiver antenna 42 m below the surface. (From Clough and Bentley, 1970.)

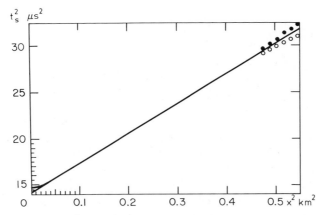

Fig. 52. (Time)^2 vs. (distance)^2 plot for an oblique reflection profile on the Ross ice shelf, showing lateral wave (o's), surface wave (dots) and apparent reflection delay at very short distances. Straight line corresponds to the linear least-square fit to the arrivals shown in Figure 53. (From Jezek, et al., 1978.)

they can cause errors to be made in the determination of mean velocities through the ice by the oblique reflection method (Figure 52).

A common occurrence (Robin et al., 1969; Clough and Bentley, 1970; Bogorodsky et al., 1970; van Autenboer and Decleir, 1971; Jiracek and Bentley, 1971) [153; 50; 205; 174; 111] which is probably related to instrumental rather than wave propagation factors, but has never been identified with certainty, is the relative delay of vertical reflections compared to the intercept time on oblique reflection profiles (Figure 53; see also Figure 40 and the discussion on pp. 102–103). Differ-

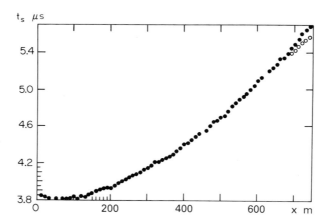

Fig. 53. Travel-time plot for the reflection profile of Figure 52, showing lateral wave (o's) and the apparent echo delay at distances of less than 50 m. The delay reaches about 0.5 μs for vertical echoes.

ences approaching 1 μs have been observed, although 0.2 μs or less is more common and sometimes the delay does not occur at all. Jezek et al. (1978) [109] believe it results from low-amplitude cycles in the rise of the transmitted pulse too weak to be observed at propagation path lengths greater than about 80 m. They never appear on the reflected signals, but are large enough to cause the initiation of the direct wave to appear increasingly early as the antennas are brought closer together. This interpretation was supported by experiments during July 1977 on the Greenland ice sheet, where photographs of the transmitted pulse showed the attenuation and loss of the initial, low-amplitude cycles.

Note added in proof: Jezek and Roeloffs (1983) [200] have recently shown that a major factor in the unexpectely high velocities found from oblique-reflection measurements is an amplitude-dependent delay in the receiver of the SPRI Mark II (and presumably also other) radar systems that have been used widely in these measurements. It is not clear yet whether this is the entire explanation.

6.4. Absorption of Electromagnetic Waves

EFFECTIVE TEMPERATURE OF GLACIERS

It was shown in Section 1.5 that glacial ice is a member of a class of substances called dielectrics. An electromagnetic wave propagating in such substances attenuates as its energy is dissipated, i.e., transformed into heat. These irreversible losses are represented by the imaginary part of the complex dielectric permittivity. However, the absorption of electromagnetic energy is usually expressed by the 'loss tangent' (tangent of the dielectric loss angle), $\tan \delta$ (1.1, 1.2). $\tan \delta$ is a very important characteristic which not only defines the loss of electromagnetic waves during their propagation through the glacier, but also makes it possible to choose the most promising wavelength ranges to be used for radar sounding in cold and temperate glaciers. Since

$$\tan \delta = \frac{\sigma(T)}{\omega \varepsilon_0 \, \varepsilon'(\rho)},$$

tan δ depends on the frequency of the electromagnetic wave, ice density, and temperature. A similar relationship holds for the snow.

Values of tan δ for antarctic ice were estimated by an indirect method in 1966 (Bogorodsky, 1978) [5]. Estimates were made using radar sounding measurements along the traverse route from Mirny observatory to Pionerskaya station. The method of estimation and the tan δ values are given in the reference. Values of tan δ are used for the evaluation of the total attenuation of radio signals during their propagation through ice. A comparison of experimental and calculated values made on the basis of 36 field measurements of losses allows us to obtain the most reliable mean values of tan δ through the whole ice sheet, including the snow/firn layer. For f_p = 213 MHz, tan δ varies from 4.2×10^{-4} to 7×10^{-4}. Measurements were made between 25 and 100 km along the Mirny-Pionerskaya route, where the mean ice temperature was about −20°C.

According to the data (Bogorodsky, 1978; Bogorodsky and Fedorov, 1967) [5, 11], the most effective frequency range for radar sounding of glaciers is from 30 to 500 MHz. A more complete set of data on tan δ for glacier ice is given by Jiracek (1967) (Table XV). The results obtained by Jiracek show a strong dependence of tan δ on temperature, density, and frequency.

The temperature of glaciers is one of the most important parameters controlling the physical conditions of glaciers. Indeed, the temperature regime of glaciers governs the main processes of their accumulation and ablation as well as their dynamic characteristics. The influence of ice temperature on the electrical properties of ice is considerable, so it should not be disregarded when designing radar equipment for ice research.

Direct measurements of glacier temperature through the entire thickness are scanty. Therefore, many scientists have used instead a rough theoretical calculation of glacial temperatures. It is impossible to take into account all the processes affecting the temperature and to work out and solve a closed system of differential equations. Nevertheless, information on englacial temperatures is necessary for the solution of various problems in the field of dynamic glaciology. Practical work is still based on theoretical results obtained by M.I. Budyko, I.A. Zotikov, P.A. Shumsky, and V.N. Bogoslovsky.

In studying glaciers as climatic indicators it is important to take into account not only marginal and surface melting, but also bottom melting. There is ample evidence indicating that zones of ice melting do occur under the Antarctic ice sheet. Reliable data on temperatures in the ice sheet and their variations in time and space are crucial in climate-related studies. The radar sounding technique helps in obtaining these data.

TABLE XV
Loss tangents

Area	Ice density (μg m^{-3})	Ice temperature (°C)	Frequency (MHz)		
			150	300	600
Iceland: glacial ice	0.818	−1	0.00220	0.00108	0.00052
		−5	0.00144	0.00076	0.00040
		−10	0.00110	0.00055	0.00028
		−20	0.00068	0.00033	0.00019
		−30	0.00043	0.00021	0.00013
		−40	0.00026	0.00013	0.00008
		−50	0.00014	0.00008	0.00005
		−60	0.00005	0.00003	0.00003
Little America, Antarctica	0.881	−1	0.00370	0.00180	0.00106
		−5	0.00260	0.00130	0.00072
		−10	0.00217	0.00108	0.00056
		−20	0.00154	0.00078	0.00038
		−30	0.00116	0.00057	0.00029
		−40	0.00085	0.00044	0.00025

Absorption losses constitute a principal part of the total losses of electromagnetic energy due to its propagation through ice (Section 1.5, Equations (1.6)–(1.8)). The losses depend on ice temperature. Thus, knowing the total losses N_Σ and having reliable data on tan δ and ice thickness, we can obtain the expression for the mean effective temperature of glaciers using (1.7) and (1.8). In this case, the technical specifications of the radar equipment are assumed to be known. The mean temperature is the average of the temperatures of different glacial layers. In principle, we can obtain the temperatures of separate layers by radar sounding, if their interfaces have electromagnetic contrasts. The mean effective temperature, also known as the absorption temperature, is one of the most important characteristics of a glacier. It can be obtained in real time over long traverses and large areas. Table XVI gives values of tan δ estimated from the data of Table XV, and values of the specific losses due to absorption (N_A) for different temperatures at a frequency of 60 MHz.

Table XVI shows a strong dependence of absorption losses on ice temperature. Thus, the mean temperature of thick layers can be precisely measured by the attenuation of a radar pulse.

In 1968–1969, a multi-frequency sounding was carried out along 175 km of the route south from Molodezhnaya station (Antarctica) (Figure 54), using radars RLS-5-67 and RLS-60-67. The results were used to plot the total attenuation of radio signals at six frequencies (Figure 55). Mean annual air temperatures and a profile of ice thickness are also shown in Figure 55. The plots show a weak dependence of the total attenuation on frequency. The effective temperature of the

SCIENTIFIC RESULTS 123

TABLE XVI

Loss tangent and specific absorption versus ice temperature

	Ice temperature (°C)							
	−1	−5	−10	−20	−30	−40	−50	−60
$\tan \delta \times 10^3$	8.8	5.8	4.4	2.7	1.7	1.0	0.6	0.2
N_A(dB km^{-1})	85.0	56.0	42.5	26.0	16.4	9.7	5.8	1.9

glacier was claculated on the basis of the total signal attenuation and the ice thickness (Figure 54).

The slope of the curve defining the dependence of electromagnetic wave absorption on temperature varies from 0.8 to 5 dB °C^{-1} for major Antarctic glacial zones. The values prove the sensitivity of this method.

The experimentally derived and calculated values of the absorption contain random and systematic errors. The random errors may be attributed partly to variations in the effective reflectivity of the bedrock surface and partly to polarization effects. The systematic errors result from the calibration inaccuracy caused by the imperfect calculation of dissipation at the surface and at the bedrock. Random

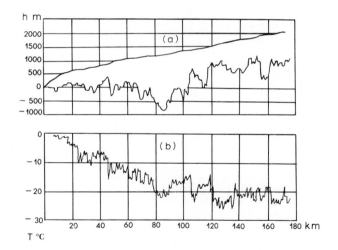

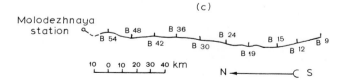

Fig. 54. Glacier profile and bedrock morphology (a) and effective glacier temperature (b) along the route (c).

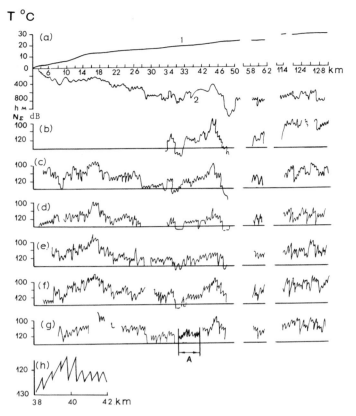

Fig. 55. Averaged total attenuation of different frequency signals along the measurement track. (a) mean annual air temperature (1) and ice thickness (2) along the track; (b)–(g) total attenuation plots for soundings at frequencies: 60 MHz (b), 70 MHz (c), 100 MHz (d), 140 MHz (e), 200 MHz (f), 440 MHz (g); (h): expanded plot of echo attenuation along section A.

errors can be minimized by averaging, e.g., by integration on the oscilloscope screen during recording. A comparison of the experimental results with the results obtained by radar measurements in holes enables a direct estimate of the errors in absorption–temperature measurements to be made. Comparison between calculated and experimental values of total attenuation shows that systematic errors vary from 3 to 5 dB. In this case the accuracy of the absorption temperature estimate varies from 0.6 to 6°C. The curve of glacier temperature as a function of distance from the ocean demonstrates a strong correlation with the corresponding variations in mean annual temperature. A coastal region, where it is difficult to calculate the losses due to melting, is an exception. Knowing the absorption temperature, average annual temperatures, and distribution of temperatures with depth, it is possible to obtain a rough estimate of ice temperature at the bed. According to numerous radar measurements near the KAM-11 hole (the Mirny area), the absorption temperature at different frequencies is about −11 to −12°C. This temperature corresponds to an ice temperature at the bed of about −5°C. A better knowledge of temperature variations with depth in the ice would make possible the determination of the temperature of any particular ice layer from the absorption temperature.

6.5. Polarization Studies

The study of the effects of glacier ice on the polarization of radio waves traveling through the ice is of substantial potential value for examining the internal and basal characteristics of glaciers and ice sheets. The depolarization of waves (transfer of energy from predicted strong polarizations to predicted weak ones) can be attributed to (1) departures from layer geometry which can occur at either the top or the bottom of a glacier, (2) inhomogeneity, and (3) anisotropy. Observation of depolarization that have been made in the field probably include examples of all three of these possibilities. The potential of polarization studies is still largely unrealized.

Measurements made in 1967 (Pasynkov, 1980) [20] revealed signal depolarization during vertical radar soundings. The equipment used in field experiments consisted of a radar transmitter with a dipole antenna as the source of linearly-polarized waves, a dipole receiving antenna, and a radar receiver. If electromagnetic waves received by the receiver are linearly polarized, then the amplitude of an output signal (U) is described by

$$U = a|\mathbf{E}| \cos \gamma, \qquad (6.5)$$

where $\mathbf{E}$ denotes the field strength at the receiving antenna, γ is the angle between the direction of vector $\mathbf{E}$ and the receiving dipole axis, and a is the receiver gain.

This equipment made it possible to obtain polarization patterns that are functions of the amplitude of the echoes from bedrock and internal layers and the angle between the transmitting and receiving dipoles. The receiving dipole was rotated through 360° in the horizontal plane by means of a rotating device that also synchronously rotated a photorecorder. The photographic record displayed the dependence of the brightness, that is, the echo amplitude, on the angle between the transmitting and receiving dipoles. Measurements were made at a frequency of 440 MHz, pulse power was 500 W, and the pulse length was 0.3 μs. The rotation axis was normal to the ice surface and passed through the middle of the dipole. Antennas were mounted 2 m above the ice surface. Measurements were made in the vicinity of Molodezhnaya Station. They indicated significant changes of the polarization plane and polarization ellipticity with an angle change of 30°. Figure 56 demonstrates sample polarization patterns. 300 of them were otained at 39 sites, where ice thickness ranged from 76 to 315 m.

The observed relationships between the angle of rotation of the polarization plane, $\theta(\alpha)$, and the azimuth angle of the receiving dipole were found to be useful

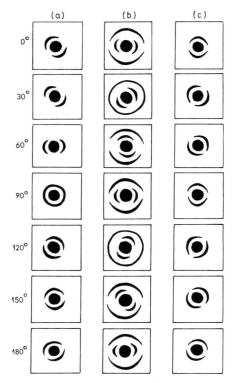

Fig. 56. Polarization diagrams at various azimuthal angles, α, of the transmitting dipole. (a) at point 4; (b) at point 5; (c) at point 7.

in glaciological studies. They appear to be almost linear in most cases, and can be divided into three categories: (1) K (slope) = 0 (Figure 57(a)); (2) $K = -1$ (Figure 57(b)); and (3) $K = -2$ (Figure 57(c)). All three groups are described by

$$\theta(\alpha) = K\alpha + A + \delta(\alpha), \qquad (6.6)$$

where K takes the discrete values 0, -1, and -2, whereas $\delta(\alpha)$ in 52 cases out of 56 is numerically less than 30°. Table XVII gives the values of the mean slopes obtained in the field.

Out of 56 echoes (44 from the bedrock and 12 from internal layers), in 9 cases $K = 0$, in 20 cases $K = -1$, in 23 cases $K = -2$, and in 4 cases the change in $\theta(\alpha)$ was too large to be classified in any of the three groups.

The Equation (6.6) with $K = 0$ corresponds to the case in which the angle between the transmitting dipole and the plane of the echo is constant on the average, within an accuracy of $\delta(\alpha)$. $K = -1$ corresponds to the case in which the orientation of the transmitting dipole changes while the echo polarization plane does not, i.e., the ice together with its reflecting boundary behaves as a polaroid. When $K = -2$, a rotation of the polarization plane of a transmitted signal by a certain angle causes the rotation of the echo polarization plane by the same angle but in the opposite direction. It was found in the field, for relationships of the second type, that when θ was close to 90°, the echoes were much weaker than with other angles. For relationships of the first and third types the echo intensity variations in the polarization diagrams were also observed, but no regular dependence of the echo strength on the angle of the polarization plane was seen.

Information on the depolarization properties of ice can be derived from the ellipticity of the echo polarization, which is observed when the azimuthal angle of the transmitting dipole changes. Thus, Figure 56(b) shows the smallest ellipticity (closest to circular) at α equal to 30° and 120°; in other cases, when the functions

TABLE XVII
Mean slopes of the angle of rotation of the polarization plane, for bedrock and layer echoes

Number of site	First echo		Second echo	
	Echo time (μs)	K	Echo time (μs)	K
1	3.7	0	2.6	−1
2	3.3	−1	—	—
3	4.7	−2	—	—
4	1.7	−1	—	—
5	3.5	0	1.3	−2
6	3.1	0	3.7	−2
7	1.3	−2	—	—
8	1.2	not determined	1.0	−2
9	1.0	−1	1.2	−2
10	1.1	−2	—	—
11	1.3	−2	—	—
12	1.7	−2	1.2	−1
13	1.6	−1	2.0	−1
14	1.8	−1	1.2	−1
15	1.8	0	—	—
16	2.0	−2	1.2	−1
17	2.0	−2	2.2	−2
18	2.0	−2	2.2	−2
19	1.6	−2	1.9	−2
20	1.7	−1	—	—
21	1.5	−2	1.0	−1
22	1.5	−2	1.0	0
23	1.5	not determined	0.8	0
24	1.6	0	—	—
25	0.8	−1	1.8	−2
26	1.8	−2	1.2	−1
27	1.8	−1	1.6	not determined
28	2.0	−2	—	—
29	2.0	−2	—	—
30	2.2	0	1.4	−1
31	3.1	−1	3.4	−1
32	3.3	−1	3.1	not determined
33	3.1	−2	3.3	not determined
34	3.3	−1	3.6	0

$\theta(\alpha)$ are of the first and third types, the ellipticity becomes minimal for a half turn of the dipole with a 90° shift.

Glacier ice thus seems to exhibit double refraction. When the difference between the ordinary and extraordinary waves ('o' and 'e') equals an even number of half-periods, no change in the radiated signal occurs. This is in agreement with (6.6), when $K = 0$ and $A = 0$ (the relationships shown in Fig. 57(a)). When the phase difference between the arriving ordinary ('o') and extraordinary waves ('e') is an odd multiple of π, a clockwise rotation in the polarization of the transmitted wave will cause the echo polarization plane to turn counter-clockwise by the same angle. This is in agreement with the $\theta(\alpha)$ function of the third type (Fig. 57(c)). When the phase difference between 'o' 'e' is not an intergral multiple of π, the general character of the dependence described by (6.6), with $K = 0$ or -2 is maintained for linear polarization only when the transmitting dipole is oriented normal to the

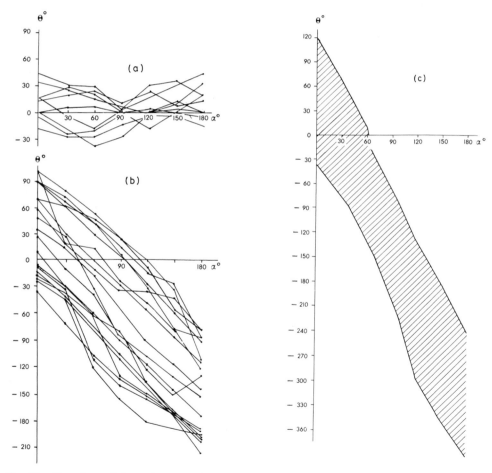

Fig. 57. Experimental determination of the dependence of the rotation angle, θ, of the plane of polarization, for bedrock and layer echoes, on the azimuthal angle, α, of transmitting dipole.

principal axes of the dielectric permittivity tensor; it reaches minimum ellipticity when it is oriented at 45° to the principal axes. This is illustrated by Figure 56(a). At several sites the $\theta(\alpha)$ relationship obtained for echoes from closely spaced reflectors correspond to different K values. It is seen from Table XVII that for K to vary from 0 to -2, the difference in echo time should be 0.5–0.7 μs.

Assuming that this corresponds to a change in the difference in arrival times of the ordinary and extraordinary waves of half a period, we can calculate the relative difference of the refraction coefficients for the ordinary and extraordinary waves (n_o and n_e):

$$\frac{|n_o - n_e|}{n_o + n_e} = \frac{1}{4f_p} = 0.0008-0.0011,$$

where f_p denotes the operating frequency of the radar.

In some cases, shown in Table XVII, rotation of the transmitting dipole does not result in a change in the polarization of the internal-layer echo, whereas the polarization of the echo from distant bedrock changes in accordance with (6.6) for $K = 0$ or $K = -2$. This can be attributed to anisotropy in the reflection coefficients for the internal layers and the bedrock. The joint effect of both phenomena (double refraction and anisotropy in the reflection coefficient) explains $\delta(\alpha)$ in (6.6), as well as the intensity change, at the maximum in the polarization diagrams with change in the azimuthal angle of the transmitting dipole, for relationships of the first and third type.

In some cases it is difficult to plot $\theta(\alpha)$, since the echo is a composite of echoes from several bedrock areas that have different anisotropies in reflection coefficient and are situated at different distances within the pulse length. If there is only a small difference in reflection coefficients between the ordinary and extraordinary waves, the phase difference between 'o' and 'e' would not change significantly within a pulse length of 0.3 μs, so the components would behave in almost the same way. That is why in many cases we can get $\theta(\alpha)$ relationships.

Double refraction can be caused either by a normal crystalline ice structure, or by ice photoelasticity in the radio range.

In the West, the first polarization work was done by Jiracek (1967) [110], who found striking evidence of anisotropy in the characteristics of the ice both at South Pole Station and on Skelton Glacier. At South Pole Station, in fact, the bottom echo could not be received when the antennas were parallel, and the echo strength appeared to be at a maximum when the antennas were oriented perpendicular to each other. This indicated that the echo amplitude maximum was rotated approximately 90° from the original transmitted pulse. On Skelton Glacier, on the other hand, the amplitude of the bottom echo was practically independent of the orientation of the receiving antenna, a phenomenon that is most easily interpreted as the transformation of linearly polarized radiation into an elliptically (nearly circular) polarized state. An example of this effect is seen in Figure 58, which shows oscilloscope photographs at a transmitter–receiver separation of 800 m. The upper picture shows the results with the antennas parallel, and the lower with the antennas perpendicular. The first signal in each photograph is the direct airwave

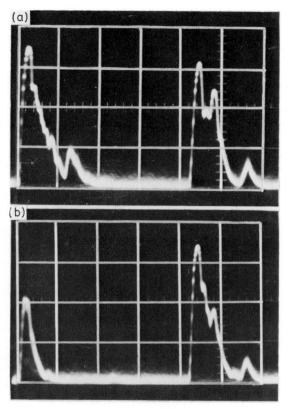

Fig. 58. A-display recordings made on the Skelton Glacier with transmitter–receiver separation of 250 m. Antennas were oriented (a) parallel and (b) perpendicular with respect to each other. The left-hand signal in each record is the direct wave, and the right-hand signal the bottom-reflected wave. The ice thickness is about 800 m. (From Jiracek, 1967.)

and the second the basal reflection. Note that the airwave is greatly diminished in amplitude, as was expected, whereas the reflection is essentially unchanged. All equipment settings were the same for both photographs.

Clough (1974, Bentley, 1975) [52, 34] extended Jiracek's work on Skelton Glacier by making measurements both on the grounded ice and on the floating ice of Skelton Inlet. On Skelton Glacier, measurements of reflection amplitude were repeated at 22.5° intervals in azimuth with successively parallel, perpendicular (cross-polarized), and collinear orientations of the sounding antennas. If elliptical polarization arising from an inclined preferred orientation of the c-axes occurs, the cross-polarization should be at a minimum when the dipoles are aligned parallel or perpendicular to the vertical plane containing the preferred orientation, and at a maximum at 45° to these positions. The experimental results suggest such a pattern relative to a plane with an orientation roughly parallel (or perpendicular) to the direction of the glacier movement. The experiment on the floating ice was conducted a few kilometers downstream. The depolarized component is much weaker, and there is no apparent dependence of amplitude on azimuthal angle.

Sharp changes in crystal orientation may produce significant reflection coefficients if the anisotropy is as high as 1% (Clough, 1974) [52]. For such a boundary the reflectivity should depend strongly upon the direction of polarization of the incident wave. A possible example of this from an internal echo in the Skelton Glacier is shown in Figure 59.

From observations in the region of Dye-3 station in central southern Greenland, Hargreaves (1977a, b) [93, 94] showed that electromagnetic waves reflected from the internal layers of the Greenland ice sheet were elliptically polarized and, by observing the change in the polarization of reflections from successively deeper layers, he obtained an estimate of the birefringence of the ice sheet. On the assumption that the optic axis of the ice was perpendicular to the direction of propagation, he found a value of 10^{-3} for the difference between the principal values of the dielectric constant. If the optic axis is at any other orientation, of course, the anisotropy must be larger. The most likely explanation for the birefringence is anisotropy of ε_∞ in monocrystalline ice, combined with a preferred orientation of the ice crystals in the ice sheet. Using evidence from other locations on the crystal fabric that might be expected at Dye-3, Hargreaves (1978) [95] concluded that the experimentally observed birefringence is consistent with reasonable fabrics and an anisotropy of the dielectric constant slightly less than 1%,

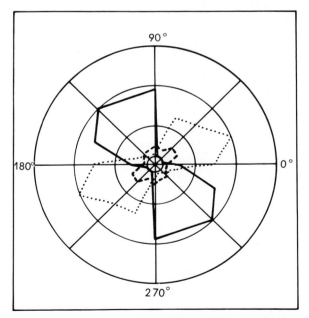

Fig. 59. Reflection amplitudes as a function of azimuth for different orientations of the transmitting and receiving antennas. Antenna orientations: parallel (solid line), perpendicular (dashed line), collinear (dotted line). The site is on Skelton Glacier, and the reflecting horizon is internal to the glacier, at a depth of about 500 m. (From Clough, 1974.)

that is, slightly less than the accuracy of the best laboratory measurements, which have thus far failed to deted any anisotropy (see Section 3.2). Further knowledge of the propagation characteristics should lead to the possibility of studying crystal orientations *in situ*. However, the rapid change in polarization phenomena in a short distance along Skelton Glacier from a region of probably high stress to one of low stress suggests that there may be a stress-induced birefringence, since the time it would take the ice to move a few kilometers down Skelton Glacier would appear to be insufficient to produce a major change in the crystalline fabric.

Experiments probably showing the depolarizing effect of scattering from inhomogeneities within the glacier were carried out on Athabasca Glacier in Alberta, Canada (Strangway *et al.*, 1974) [166]. Little evidence of depolarization was found at 1, 2, and 4 MHz, but at 8 MHz the depolarization was large, and at 16 and 32 MHz the parallel and perpendicular components showed equal amplitudes. In the case of Athabasca Glacier, the effect of the bottom could be ruled out by the strong frequency dependence and the fact that the depolarization was observed also in waves propagating through the ice just beneath the surface. Surface roughness was discounted because of the absence of strong depolarization in a similar experiment carried out on the Moon, the surface of which is nearly as rough as that of Athabasca Glacier. Bulk anisotropy can also be ruled out by the strong frequency dependence. Watts and England (1976) [186] show that a model made up of spherical scatters with a Gaussian distribution of radii about a mean of 1 m with a standard deviation of 0.2 to 0.4 m explains the observations on Athabasca Glacier.

Experiments conducted at several station on the Ross ice shelf in the course of the RIGGS program showed evidence of depolarization at only one centrally located site (M14) where a layer of saline ice 10–14 m thick was observed by radar profiling (S. Shabtaie, personal communication, 1979). There, elliptical polarization of the bottom echo was indicated by strong cross-polarization that showed two maxima oriented in different azimuths. Additionally, the polarization measurement showed dependence of echo strength on azimuthal angle. Both phenomena are probably due to the basal sea-ice layer, which, if it formed in the presence of a current, would have a strong preferred orientation of c-axes aligned parallel to the current in the horizontal plane (Weeks and Gow, 1980) [189]. Azimuthal dependence of reflection strength has been observed elsewhere on oriented sea-ice (Kovacs and Morey, 1978 [120]; 1980 [121]), and the effect of an anisotropic distribution of brine channels in the ice could add to the depolarization resulting from the bulk crystalline fabric.

6.6. Ice Thickness and Subglacial Topography

The most widespread application of radar sounding is the measurement of glacier thickness. In the last 20 years the technique has been used extensively all over the world for this purpose, measurements being made both from aircraft and from surface vehicles on the Antarctic and Greenland ice sheets, on Arctic ice domes, and on mountain glaciers. It is beyond the scope of this book to attempt to review all of these results in detail. We will summarize here the areas of covesage and the more general aspects of the geographical results.

AIRCRAFT SURVEYS

Radar sounding equipment designed for airborne measurements were reviewed in Chapter 4. Airborne sounding of antarctic and arctic glaciers in Soviet expeditions includes intercalibration of all flight altimeters and ice-thickness-measuring radars, flight parameter recording, radar sounding of ice thickness, and processing of navigation and radar measurements (Bogorodsky, Trepov, Sheremetyev, 1979 [10]). In addition to the radars, there are radio altimeters recording flight height above the ice surface and barometric altimeters for measuring the altitude above sea level as well. Each radar survey is preceded by a test flight over the sea. In this flight all radio altimeter readings are recorded, as well as those of barometric altimeters and the radar. Air temperatures are also measured. Simultaneously, aerial photography is carried out in order to reveal when the aircraft flies over icebergs. Test flights are made at altitudes ranging from 100 m to 3500 m so that the scales of all recorders can be calibrated by a radio altimeter.

The navigation on inland antarctic flights is rather complicated due to the absence of any reference points. For that reason Doppler systems are used in radar surveys to measure aircraft velocity and angle of drift. If reference points such as a coastline or nunataks are met on the way, aerial photography is carried out. Radar sounding consists of recording the amplitude and intensity displays of an ice-thickness-measuring radar and the radio-altimeter display. Airborne radar measurements of ice thickness were made not only on ice sheets but on outlet glaciers and ice shelves as well. To measure the ice thickness of coastal glaciers of irregular topography with well-marked reference points in sight, a simplified survey method is employed, using visual navigation.

ICE THICKNESS MEASUREMENTS ON TRACTOR-SLEDGE TRAVERSES IN THE SOVIET ARCTIC AND ANTARCTIC EXPEDITIONS

Ice thickness measurements are needed for detailed surveys of sub-glacial topography and for comparison with glaciologic, geographic, and geodetic measurements. These measurements are also important for the study of fluctuations in the strength of bedrock echoes. The display time sweep in surface measurements is calibrated in such a way that the sweep length corresponds to 500, 1000, 2000, and 4000 m of ice thickness. The strip chart is drawn forward discretely every 2.5 m or 10 m of the track. This recording mode provides for a high accuracy of point-for-point correspondence between the radar and bedrock profiles, thus simplifying the data processing. During inland traverses measurements were made in the vicinity of Molodezhnaya (several tracks), along the route from Mirny to Vostok, and others. Figure 60 shows mainly the flight lines over Antarctica along which radar sounding of ice thicknesses were made, but, in addition, surface radar measurements are also indicated. Figure 61 presents the most characteristic segments of radar ice thickness measurements on antarctic glaciers made by the Soviet specialists.

Radio echo sounding made by U.S. and Soviet specialists indicate that, when sounding polar ice from aircraft, separate echoes from the upper and lower surfaces

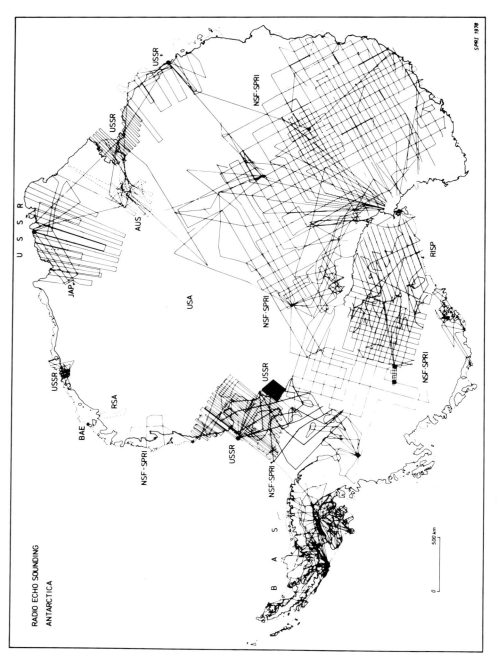

Fig. 60. Map of radar-sounding flight lines in Antarctica, as of 1978.

SCIENTIFIC RESULTS

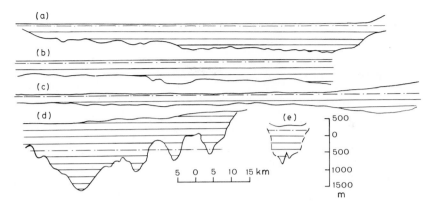

Fig. 61. Radar sounding sections of some Antarctic glaciers: a – Amery ice shelf at the boundary with Lambert Glacier (from 69°45′S, 73°43′E, to 70°43′S, 65°45′E); b – Amery ice shelf (from 68°58′S, 72°27′E, to 70°20′S, 70°21′E); c – Lazarev ice shelf (from 69°30′S, 14°22′E, to 70°29′S, 14°22′E); d – cross-section of Hays outlet glacier, 50 km from the coast (from 67°57′S, 46°64′E, to 67°59′S, 47°19′E); e – cross-section of Hays outlet glacier 5 km from the coast (from 67°40.5′S, 46°01′E, to 67°04′S, 46°21′E). Dashed line denotes sea level.

of the ice sheet are usually obtained without difficulty. The upper surface echo is normally very strong and shows little fading, whereas the strength of the lower surface echo usually varies rapidly as the observer moves. However, this fading is so rapid that the bedrock echo is virtually continuous, provided that its strength exceeds the noise level.

Owing to the continuity of bedrock echoes, few mistakes in their interpretation are likely, although occasionally an internal echo has been mistaken for an echo from bedrock. In general, the reliability of interpretation is such that radar records provide a standard against which earlier seismic and gravity depth estimates may be tested for gross errors (Drewry, 1975) [69].

Since a radar record shows the range to the nearest ice–rock interface, and not the vertical depth of ice, it does not present a true vertical profile of the bedrock surface. The same problem exists in marine sounding, but analysis is more complicated with airborne radar sounding owing to the effect of the air–ice interface. Harrison (1970) [96] and Macheret and Luchininov [197] have shown how to deal with the problem of the use of deconvolution techniques. When dealing with bottom slopes of less than, say, 1 in 20, deconvolution is of minor importance, but for steeper slopes it is essential and its use should be more widespread.

ANTARCTICA

Certainly the most dramatic application of radar thickness sounding has been in Antarctica, if for no other reason than that vast continent represents over 90% of all the glacier ice in the world. A map showing all the radar-sounding flight lines as of 1978 is shown in Figure 60. Since radar-sounding flights continue, the coverage will be more extensive by the time of the publication of this book, but the map

136 CHAPTER 6

shown gives a good general idea of the extensive amount of surveying that has been done, keeping in mind the vast extent of the Antarctic continent – roughly equal to that of all of the R.S.F.S.R., west of the Lena River.

THE INTERIOR ICE SHEET OF ANTARCTICA

The first measurements of ice thickness by radar reflections on the thick ice of interior Antarctica were made by Jiracek and Bentley in January, 1965 at South Pole Station (Jiracek, 1967) [110]. The calculated ice thickness, 2800 m, was in excellent agreement with seismic reflection data. The success at South Pole led to the inclusion of radar sounding equipment on the ground traverse in Queen Maud Land the following season. The travel time was recorded every 0.4 km and photographs were obtained at 1.6 km intervals. Echoes were received from depths as

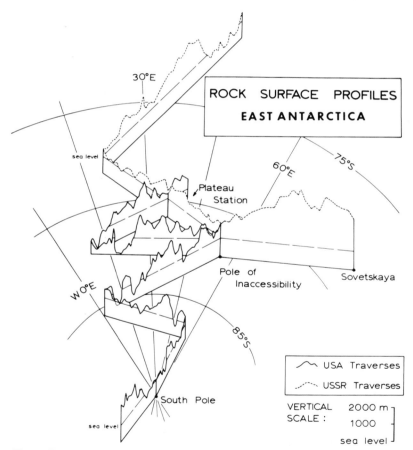

Fig. 62. Fence diagram of ice thickness along U.S. and U.S.S.R. traverses in Queen Maud Land, Antarctica. The sections along the U.S. traverse from the Pole of Relative Inaccessibility to Plateau Station was measured by radar; the other sections from seismic and gravity sounding. (From Beitzel, 1971.)

great as 3500 m and were successful over 90% of the total profile that extended for 1000 km. The soundings were continued on the final leg of the Queen Maud Land traverse in 1967 and 1968, although, for reasons not certain, bottom echoes then were successfully recorded only over about one-third of the track.

The rock surface (Figure 62) displays a striking grain which trends north-northeast and, although the contours are based on widely separated lines of data, the reality of the trend is supported by the persistence of the gross topographic features between sounding lines. The dominant feature is a massive central block, 250–300 km wide, flanked on the west by a 90-km wide valley that reaches below sea level. More than 2 km of relief is found on the west flank of the central high, comparable with the relief in the Basin and Range province of the western United States. The radar sounding data indicate that the surface of the central block is dissected by relief of the order of hundreds of meters occurring over a few kilometers.

The airborne radar program carried out jointly by the Scott Polar Research Institute and the U.S. National Science Foundation commenced with 94 hours of flying in 1967–68. Since then, a substantial part of the Antarctic has been covered, as can be seen in Figure 60.

The region between the Transantarctic Mountains and the South Pole was investigated in some detail by Drewry (1971) [66] – a typical profile is shown in Figure 63 and a map of the subglacial topography in Figure 64. The subglacial topography is complex to about latitude 88°S, with ice thicknesses ranging between 500 and 3000 m covering irregular mountainous land surface with relief of 500 to 1000 m. Farther south the relief becomes less varied.

Another area of detailed investigation within the subglacial Transantarctic Mountains lies inland of the McMurdo ice-free valleys. Sounding was carried out along subparallel lines, 10 to 15 km apart, between the mountains and about longitude 156°E (Calkin, 1971 [45]; Drewry, 1982 [72]). The map (Figure 65) shows that the mountainous terrain exposed in southern Victoria Land continues under the ice. There is a general decline in elevation inland from about 1500 m at the edge of the ice sheet to about 700 m at 156°E. Several deep valleys are cut into this surface, separating upland massifs. Four broad valleys trend directly inland while another feeds into Mulock Glacier. An exceptionally deep bedrock trough extends from 25 km inland through 'The Portal', and hence into Skelton Névé.

Another early result of the flight program was a map of the subglacial contours

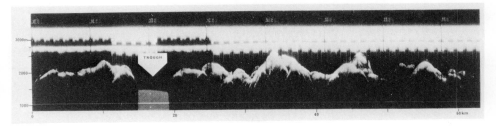

Fig. 63. Radar sounding profile of a section from the Queen Maud Mountains toward the South Pole. (From Drewry, 1971.)

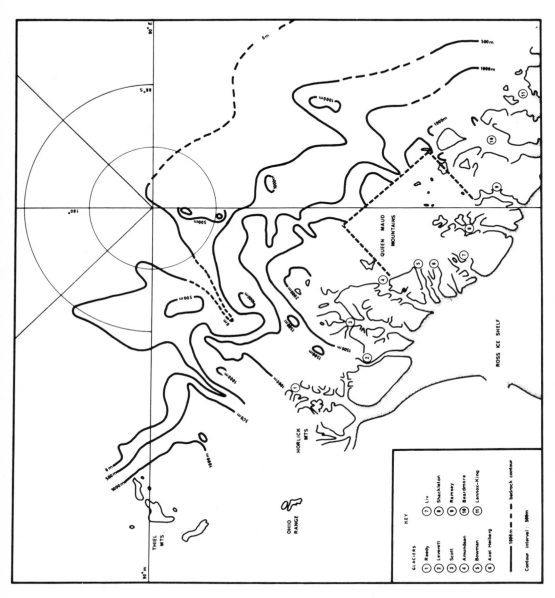

Fig. 64. Map of the subglacial topography between the Transantarctic Mountains and the South Pole. Glaciers: 1 – Reedy; 2 – Leverett; 3 – Scott; 4 –

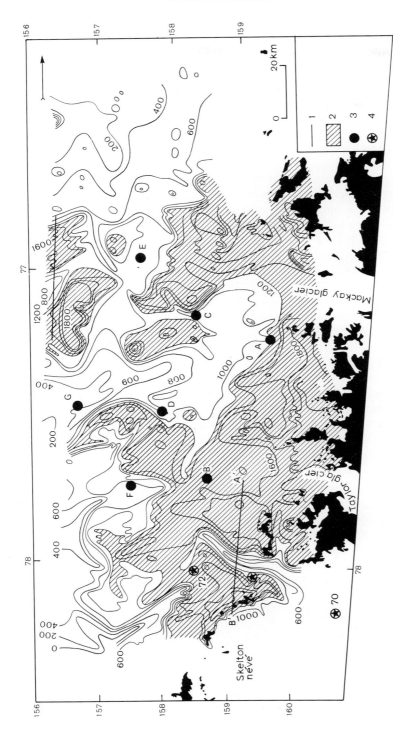

Fig. 65. Map of subglacial topography inland of McMurdo Sound. 1 – contours; contour interval 200 m. 2 – subglacial terrain elevations greater than 1200 m above sea level; 3 – subglacial water; 4 – seismic site. (From Drewry, 1982.)

around the Gamburtsev Mountains beneath the highest part of the central dome of the East Antarctic ice sheet (Robin, 1971) [149]. The Gamburtsev Mountains, an extensive subglacial mountain range, run westward from Vostok Station, and reach a maximum measured altitude of 2900 m above sea level between Sovietskaya and the Pole of Relative Inaccessibility. In contrast with the Transantarctic Mountains, the Gamburtsev Mts display an irregular upland topography with relief of the order of 1500 to 2000 m (Figure 66). The region to the east, nearer to Vostok, is characterized by subdued plateaus dissected by steep-sided valleys (Evans et al., 1972) [77]. In northern Victoria Land, variations in bottom elevation reach 1000 m over 10 km, although they appear to be related to occasional peaks rather than to an extensive mountain range. Measurements from Soviet flights around the same time showed relief up to 1500 m in Enderby Land. In contrast to these mountainous regions, much of the interior of East Antarctica forms a gently undulating lowland, at places below sea level, with ice thicknesses reaching a measured maximum of 4540 m at 77.5°S, 117.5°E (Evans and Robin, 1972) [78].

The most extensive map so far published is that covering about 2×10^6 km^2 of eastern Wilkes Land, published by Drewry (1975) [199]. Further flights in the same area have led to an updated version of that map (Steed and Drewry, 1982) [165]. The map was produced from 11 radar flights conducted during the 1971–72 season and 12 flights from 1974–75, and covers an area of 900 000 km^2 north of a line from Minna Bluff to the Sabrina Coast. The flight-line grid has a mean separation of 50 km, which is sufficiently close to identify any large-scale topographic elements, although small-scale features may often remain undetected. Where the bedrock echo was not detected, minimum values were assigned to the ice thickness.

The topography (Figure 67) shows a general north–south trend. The Transantarctic Mountains slope westward into the large, parallel-trending, Wilkes Subglacial Basin, which deepens towards the north. The basin is essentially low-lying and smooth, but contains a high area in the northwest. A more complex area, containing mountainous regions extending inland from the Adelie Coast, lies to the northwest of Wilkes Basin. This area exhibits a deep asymmetrical trench with a massive escarpment dipping west. Additional highland regions rise between the Wilkes Basin and the 'Aurora Basin' in the southern part of the area, gradually breaking into a plain to the northwest. Mertz and Ninnis Glaciers fill major fjords within the massifs.

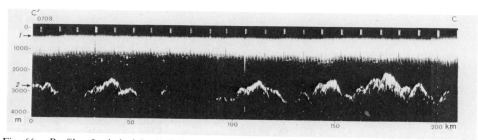

Fig. 66. Profile of subglacial relief along a section over the Gamburtsev Subglacial Mountains. The section runs southeastward from a point (C) about 100 km south-east of Sovietskaya. (From Robin, 1971.)

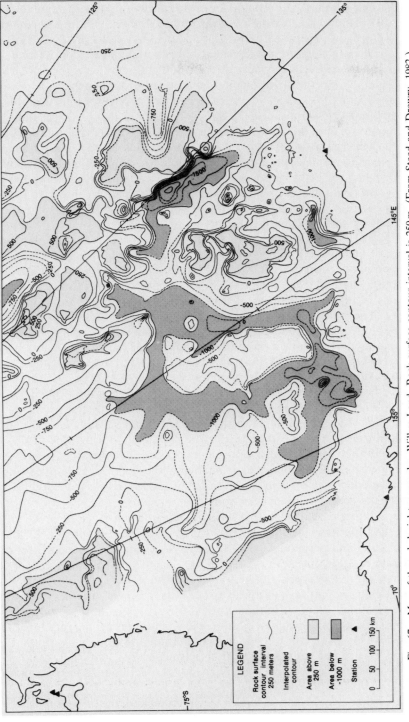

Fig. 67. Map of the subglacial topography, Wilkes Land. Rock surface contour interval is 250 m. (From Steed and Drewry, 1982.)

A profile has recently been completed from Mirny to Dome C along the surface (Young, 1979 [196]). For 650 km westward from Dome C, the profile shows relatively smooth bedrock surface, most of it about 1 km below sea level. This simple picture is broken by an abrupt steep-sided ridge 1 km high between 500 and 550 km from Dome C. Farther to the west the bedrock gradually rises to well above sea level, presumably on the flank of the Gamburtsev Mountains.

A detailed radar sounding map of the bedrock topography in the immediate vicinity of Dome C (Shabtaie et al., 1980) [162] indicates that the ice is underlain by a mountainous terrain, including a drop in one area of 500 m over 2 km. The camp is situated over a ridge whose crest is 250 m below sea level; within 11 km of the Dome C camp the land slopes away to the southwest to 350 m below sea level and to the northeast to 750 m below sea level. The local survey of Dome C was extended by airborne measurements over a 50 km box pattern flown at 10 km line spacing in each direction. The airborne survey shows a sharp ridge southeast of Dome C with a relief of nearly 1000 m.

Radar profiling from surface traverses has been carried out by the Japanese on the Mizuho Plateau south of Syowa Station. Soundings were made at 10 points on the traverse between Syowa and South Pole in 1968–69 (Yoshino and Eto, 1971 [195]). Radar sounding was conducted intermittently at intervals of 2–5 km along the traverse route from Syowa to 72°S and then in a loop westward to the Yamato Mountains. The measurements on the South Pole traverse show a bottom elevation varying by a few hundred meters above and below sea level. The measurements on the continuous profile show a mountainous subglacial relief. The bedrock surface from the coast eastward for 120 km remains fairly close to sea level with a couple of depressions several hundred meters below sea level. From the 120 km point to 380 km the bedrock surface is hilly with a very deep depression 1760 m below sea level 14 km directly inland from Shirase Glacier, presumably the site of the feeding ice stream. The general level of the bedrock surface decreases northward. A deconvolution technique has been applied to the same data by Mae (1978) [124]. This has smoothed out the bottom profile but has not changed the general characteristic, although it has diminished dramatically the depth of the valley upstream from Shirase Glacier to only about 200 m below sea level, or a few hundred meters lower than the surrounding glacier floor.

In West Antarctica, flights in 1974–75 covered an extensive area between 135°W longitude and the Ross ice shelf on a 50 km^2 grid (Rose, 1982) [157]. Flights were aligned parallel and perpendicular to the 135°W meridian, so that one set of lines approximate ice flow lines.

The map of the bedrock topography is shown Figure 68. The rock surface changes in character from smooth near the Ross ice shelf, where the ice streams follow well-defined channels in the bedrock, to very rough in the southeast. South of the Siple Coast a major channel, 50 km wide and dropping in depth to more than 1750 m below sea level, winds towards Byrd Station and connects with the Bentley Subglacial Trench (already known from seismic work), which is seen at 80°S, 110°W. Two other deep channels are aligned with the Shimuzu ice stream (2060 m below sea level) and Reedy Glacier.

Further inland ice stream channels degenerate and the topography becomes

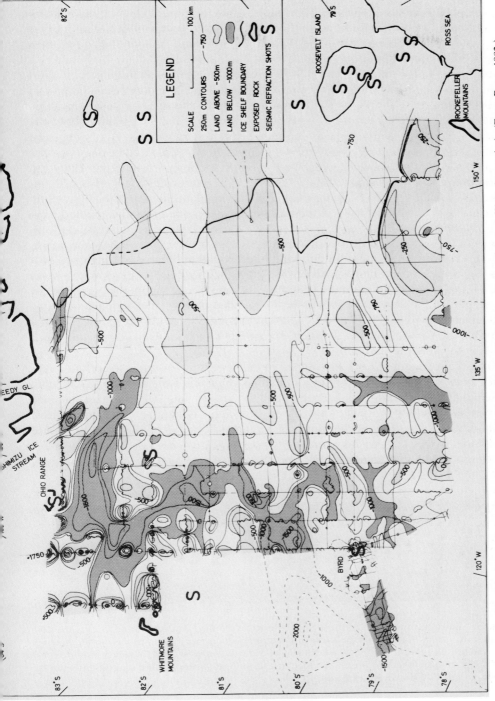

Fig. 68. Map of the subglacial topography beneath the Rockefeller Plateau (western West Antarctica). (From Rose, 1982.)

progressively rougher. Around the Whitmore and Thiel Mountains the topography is the roughest in the area; large blocks of heavily dissected mountains are divided by steep-sided valleys, typically 20 km wide and 2 km deep. In general, the bedrock is well below sea level. Removal of the ice sheet and isostatic uplift would produce only a few islands principally in the Whitmore Mountains area but also to the north and west of Byrd Station.

Airborne radar sounding in 1977–79 covered the southeastern quarter of West Antarctica, and a section between Byrd Station and the Amundsen Sea (Jankowski and Drewry, 1981 [106]). A new feature revealed by the survey is a sinuous ridge running across the center of the Byrd Subglacial Basin, standing 1000 m above the floor of the basin in some places. At the northern margin of the Byrd Subglacial Basin the outcropping volcanoes are steep-sided and are separated by deep, narrow troughs. A major subglacial highland extends southwards from the Ellsworth Mountains. The subglacial terrain is extremely rugged, possessing a relief of 2 km. There is a well-defined boundary between the highland area and the Byrd Subglacial Basin. The Whitmore Mountains form a steep-sided massif separated from the Ellsworth Mountains highland area by a narrow trough. Other scattered outcrops to the south of the Ellsworth Mountains appear to be part of the same mountainous region. The Transantarctic Mountains constitute a separate highland area and form a semi-continuous chain across Antarctica. Two deep valleys almost sever the Horlick Mountains from the rest of the Transantarctic Mountains, whereas the main part of the Pensacola Mountains is bounded on two sides by deep, narrow, troughs occupied by the Foundation ice stream and Support Force Glacier. At the head of the Ronne/Filchner ice shelf, the bedrock forms a deep depression which widens towards the ice shelf.

The radar sounding work has helped to delineate ice flow lines. Near the boundary between the West Antarctic inland ice and the Ross ice shelf, the main discharge of ice takes place through five major ice streams. The southern-most ice stream is largely a continuation of Reedy Glacier, discharging ice from the East Antarctic plateau. The northernmost ice stream is bounded on the north by the ice of Marie Byrd Land. In between the ice streams are domes or ridges of slowly-moving ice which gradually merge into the inland ice sheet of West Antarctica.

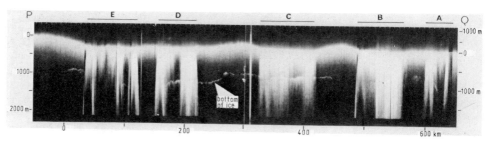

Fig. 69. Radar record along a north–south profile across the West Antarctic inland ice near the Ross ice shelf. Letters A to E show the location of the corresponding ice streams on this profile. Note the different scales of elevation in air and depth in ice due to different velocities of radio waves in ice and air. The surface profile is correctly recorded, but the differing scales produce some distortion of the bottom relief. (From Robin et al., 1970.)

The evidence for these conclusions is shown in a north–south profile across the inland ice near the Ross ice shelf (Figure 69). The fast-moving ice streams are heavily crevassed, particularly near their margins, so that surface echoes from the ice streams have a long duration and are spread across most of a film, often obscuring the bottom echo. The ice domes between the ice streams show up clearly in the surface profile and in the absence of crevassing (Robin et al., 1970) [155].

COASTAL AREAS OF THE INLAND ICE

Studies in the coastal areas of Antarctica include the very first radar sounding made (Waite (1959) [178]; Waite and Schmidt (1962) [181]) on ice 150 m thick near the ice sheet margin in Wilkes Land. Following studies inland from Mirny made by the Soviet Antarctic Expedition in 1966, an airborne survey of the Antarctic Peninsula area was begun by the British Antarctic Survey (BAS) in 1966–67 (Swithinbank, 1968) [167]. Results from those flights and from additional flights in 1969–70 (Smith, 1972) [163], show that the subglacial topography of the Antarctic Peninsula plateau is essentially a pre-glacial surface at an elevation of about 1200 m, but with substantial relief which was either pre-existing or carved by glacial erosion. Thicknesses up to 1630 m have been recorded.

The subglacial topographies beneath Graham Land and Palmer Land differ considerably in character. The spine of Graham Land is a fairly flat rock plateau, whereas the surface beneath the snow plateau of northern Palmer Land is very uneven with slopes rising more than 1 km in a distance of 2 km (Wyeth, 1977) [194]. The measured ice thickness in Graham Land is only about 700 m. The work on the Antarctic Peninsula has continued, as can be seen by the large number of flights indicated in Figure 60.

Extensive work carried out by BAS between the Antarctic Peninsula and the main West Antarctic inland ice, has been summarized by Swithinbank (1977) [169]. The principal feature of the bedrock elevation in this area is that much of it is not only below sea level but substantially below the level of the Antarctic continental shelf. The subglacial Antarctic Peninsula seems to be separated from the Ellsworth Mountains continental fragment by deep troughs now occupied by major ice streams. There is little doubt that, as had previously been inferred from seismic sounding and gravity observations, there is potentially a channel well below sea level that connects the Bellingshausen Sea with the Weddell Sea. The most remarkable feature can be inferred from a sounding of 1860 m on the adjacent ice shelf: within 60 km of the highest mountain in Antarctica (Vincent Massif, 5140 m) there must be a trench extending to at least 1600 m below sea level.

Other Antarctic coastal studies (exclusive of ice shelves, which are considered in the next section) can be reviewed clockwise around the coastline. An early profile by Bailey and Evans (1968) [29] was completed along a single surface track inland of the Brunt ice shelf in 1965. Eastward from the edge of the ice shelf, the bedrock surface has the appearance of a plain 100–150 m below sea level, the deeper parts probably representing valleys which cut across the route. The valleys may have been cut by glacial erosion, but if allowance for isostatic rebound after removal of the ice load is made, the plain would be well above sea level, so that the valleys could also have been cut by stream action. Along the main part of the route inland, the existing rock is generally below sea level up to a point where the mountains rise steeply to an elevation of 500 m as the Theron Mountains are approached.

Soundings south of Sanae Station were carried out by the South African Antarctic Research Expedition (Schaefer, 1972 [159]; 1973 [160]; Van Zyl, 1973 [175]). Profiles are given in the cited works, but no map has yet been produced. Subglacial topography is rugged between the mountains of the Ahlmann-Ruggen, with depths reaching 1000 m below sea level beneath the Viddalen. A particularly interesting portion of the work was the good general agreement between ice depths measured by radar and those obtained seismically by Robin (1958) [147] (see Section 6.1).

Aerial sounding in the Lambert Glacier basin was carried out by the Australian National Antarctic Research Expedition (ANARE) between 1971–72 and 1973–74. The maximum ice thickness measured was about 2500 m in the middle of the Lambert Glacier, at the confluence of three feeding ice streams (Morgan and Budd, 1975 [128]). Other zones more than 2000 m deep included the centers of the valleys of the three ice streams further upstream. Still further inland, the ice which flows from the western side appears to be some 2000 m thick, whereas that flowing in from the east is closer to 1500 m thickness. This picture is broken by the deep valleys and ridges of over 1000 m relief which channel the ice into the main Lambert Glacier valley.

Below the point where the feeding glaciers join it, the Lambert Glacier is remarkably uniform, its thickness diminishing gradually from about 1000 m to about 800 m where it merges with the Amery ice shelf. Across the Amery ice shelf, the ice thins from 800 m to about 200 m at the front. Extensive soundings have been made on the Amery ice shelf, but the report on that work is not yet available.

The most remarkable feature of the radar-sounding results is the very deep bedrock depression near the confluence of Lambert Glacier and its tributaries (Figure 70). Deep hollows are often associated with the confluence of glaciers, but the depression in the Lambert Glacier is on a much greater scale. The echo strength was particularly marked across the deep section, even stronger and clearer than in neighboring shallower regions. For about 15 km the section also showed a remarkably flat base. These considerations suggest the possibility of a basal meltwater lake existing in the depression, similar to lakes found below the Antarctic ice sheet by Oswald and Robin (1973) [135]. The increased echo strength is similar to the effect they observed. Along one measured profile the bedrock slope into the depression is more than 10 times the surface slope of the ice, sufficient to cut off basal water flow. If this were to occur all the way around the depression, it would be sufficient to cause a meltwater lake. In addition, temperature calculations give melting-point values for the bottom of the depression. Thus the existence of the basal meltwater lake seems highly likely.

ANARE has also been conducting an extensive program of radar mapping in Enderby Land, continuing at least through the 1979–80 field season. The aim is to cover the region on a 20 km grid spacing. That work has been supplemented by measurements along a ground traverse that followed the 2000 m surface elevation contour from south of Mawson to near Sandercock Nunataks, thus meeting the

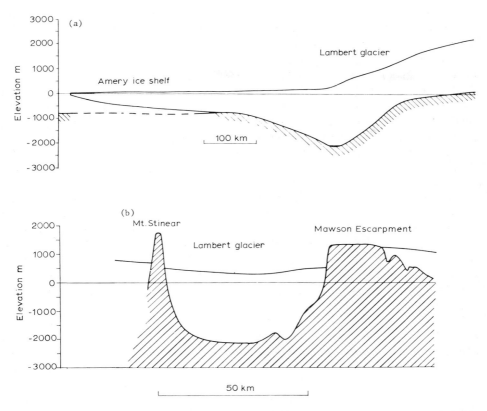

Fig. 70. Longitudinal (a) and transverse (b) sections, Lambert Glacier. (From Morgan and Budd, 1975.)

eastern-most extension of the Japanese work. From the first results of the airborne radar sounding program and the other available data, the main features of the bedrock map can be described. There is a general southward increase in bedrock elevation and a deep below-sea-level trench exists between Edward VIII Gulf and Amundsen Bay, which isolates the Napier and Tula Mountains on the dome of Enderby Land. Wilma and Robert Glaciers and Beaver Glacier, respectively, occupy the ends of this trench. A second trench cutting off the Scott and Raggatt Mountains from the area to the east runs south-southwest from the Edward VIII – Amundsen trench to link with a valley following the line of the Thyer Glacier to Casey Bay. These trenches may be the expression of deep-seated geological boundaries (Allison et al., 1982) [26].

A profile has been run along the surface from the summit of Law Dome to a point 140 km inland towards Vostok (Budd and Young, 1979) [44]. The profile shows very clearly the deep subglacial trough just south of Law Dome that connects to valleys beneath either Vanderford or Totten Glacier, or both. Southward from

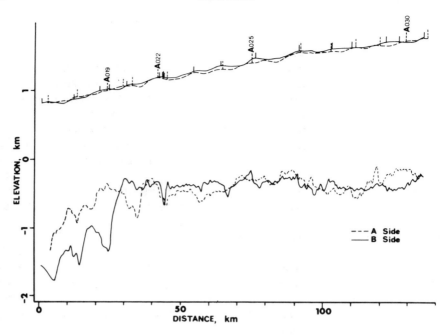

Fig. 71. Detailed 140-km long parallel sections along two tracks 10 km apart running southward from the valley south of Law Dome, Budd Coast, Wilkes Land. (From Budd and Young, 1979.)

the margin of the trough the average level of the moderately dissected bedrock is remarkably constant at about 400 m below sea level (Figure 71).

High resolution, monopulse radar systems, although designed principally for detailed study of shallow layers and reflections within the ice, can, nevertheless, be effective in measuring total ice thickness if that thickness is less than about 500 m. Kovacs (1980) [116] showed the steep subglacial face of a nunatak on the flank of the Transantarctic Mountains facing the East Antarctic ice plateau. He also showed from a detailed profile about 1 km long, on Ross Island near McMurdo Station, that depths as small at 10 m could be sounded. For other uses of the monopulse radar see Section 6.8.

THE ICE SHELVES

The Antarctic ice shelves are in some ways particularly well suited to radar studies. In part, this reflects the difficulty in measuring the thickness of floating ice by seismic techniques, as an ice–water boundary does not form a particularly good acoustic reflector. The situation is, of course, very different for electromagnetic waves. The strong reflection to be expected at a specular boundary between freshwater ice and sea water makes work on ice shelves, in principle, relatively simple. Various factors intervene to make the practice substantially different from the principle – these factors are discussed in other sections of this book. In many places, however, the echoes are indeed very strong, as simple theory would predict, and profiling of the ice shelves is effective.

Some of the first radar sounding was done on ice shelves. Early work by Waite was carried out both on the surface and from helicopters over the Ross ice shelf and the adjacent ice shelves of Marie Byrd Land (Waite, 1962 [179]). A flight over a short section of the Ross ice shelf near Kainan Bay showed the thinning of the ice shelf as the barrier was approached. On other flights the marginal ice thickness gradient was measured and a grounded subglacial ridge discovered in the ice shelf fringing the Ruppert Coast, a short section of the Getz ice shelf grounding line was delineated, and a new ice-covered island was discovered and profiled. Unfortunately very little of this work was published.

Walford (1964) [182] reported on the first traverse study of ice thickness, in 1964, in this case on the Brunt ice shelf. Additional measurements were made by Bailey in 1965 leading to an additional profile across part of the ice shelf (Bailey and Evans, 1968 [29]), but the coverage has not yet been extensive enough to lead to an ice thickness map of the ice shelf.

In the same season, Jiracek's (1967) [110] measurements were made on the Ross ice shelf at Roosevelt Island, in ice of the 'McMurdo' ice shelf, and in Skelton Inlet. This [5] work was not designed for mapping ice thickness [4], although the profiles did show an interesting rapid increase in the McMurdo ice shelf when passing laterally opposite the glaciers falling from Mounts Terra Nova and Terror, and also a rapid decrease in ice thickness of the Skelton Glacier ice just downstream of the grounding line where the width of the confining valley rapidly increases.

The work of the British Antarctic Survey in the Antarctic peninsula area has included extensive sounding of the fringing ice shelves (Swithinbank, 1968 [167]; Smith 1972 [163]). Ice thickness maps have been produced of the Larsen ice shelf (Figure 72), the Wordie ice shelf, and the shelf covering George VI Sound. In common with many other measurements on ice shelves, many of the signals were strongly attenuated due to a layer of brine-soaked firn where brine has infiltrated the ice shelf at sea level.

Additional measurements by the British Antarctic Survey were made on the northwestern part of the Ronne ice shelf (Swithinbank, 1977) [169]. Here there are three long fjord-like inlets draining the inland ice between the Ellsworth Mountains and the Antarctic Peninsula. Ice thickness increases rapidly upstream in these inlets, reaching a maximum of 1860 m on Rutford ice stream only 17 km from the eastern escarpment of the Ellsworth Mountains. This is the thickest ice ever found floating on the sea, and is nearly twice as thick as the thickest ice found on the Ross ice shelf. The rate of thinning of the ice in the inlets as one moves seaward is, evidently, principally controlled by the boundary configuration in each inlet. Ice moving down the northern ice stream thins from 1400 m to 800 m over a distance of only 45 km over which the sides of the inlet begin to diverge. On the northernmost part of the ice shelves, that is north of Korff ice rise, the shelf thins smoothly from 700 m to 300 m towards the ice front. In the constricted areas between the ice streams and Korff ice rise, however, the ice shelf remains over 1000 m thick.

Van Autenboer and Decleir (1971) [174] carried out airborne radio soundings on a grid pattern over several parts of the Fimbul ice shelf. They found that in the

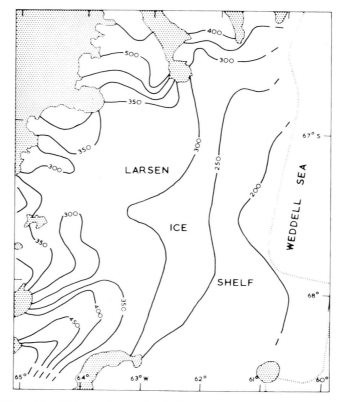

Fig. 72. Map of ice thickness, Larsen ice shelf, Antarctic Peninsula. (From Smith, 1972.)

central part of the ice shelf the ice thickness is constant at 300–350 m over a large area. Characteristic of the seaward edge of the ice shelf are 10–15 m deep depressions, which form the landward extension of many small bays in the area. Ice thickness decreases regularly towards the front in these regions with a minimum thickness of about 100 m. Towards the inner boundary of the ice shelf the thickness increases to a maximum of about 500 m. The eastern margin of the Fimbul ice shelf is formed by a major, and very fast moving, ice stream that represents the continuation of the Jutul ice stream. Limited measurements on the Jelbart ice shelf also show an average thickness of around 300–350 m.

The Blåskimen ice rise limits the Fimbul ice shelf to the northwest. Its surface relief shows a typically symmetrical dome approximately 400 m high. The ice here is approximately 500 m thick, but the subglacial topography does not show the symmetry of the surface dome. The southwestern limit of Fimbul ice shelf is formed by an elongated north–south oriented dome, the long axis of which continues to a point close to the Blåskimen ice rise. Radar sounding flights confirm that this is an extension of the inland ice sheet. The surface measurements (Schaefer, 1972 [159]; 1973 [160]; Van Zyl, 1973 [175]; Barnard, 1975 [31]) have shown good agreement with the earlier results from airborne surveys.

Extensive surveying of the Amery ice shelf region has been carried out by Soviet Antarctic Expeditions. Soviet Antarctic Expeditions have also worked extensively on Novolazarevskaya ice shelf and on Filchner ice shelf.

Extensive surveying has been carried out, over a number of years, on the Ross ice shelf. The radar sounding on the Ross ice shelf has been carried out both from the air as part of the SPRI/NSF sounding program and from a combined light aircraft and surface measurement program undertaken by the Ross Ice Shelf Geophysical and Glaciological Survey (RIGGS). The first map to be published was that presented by Robin (1975) [199]. Robin's map was drawn on the basis of some 35 000 km of sounding track. In producing the map, navigational inaccuracies were the chief difficulties. Much of the work was done before the installation of an intertial navigation system in the sounding aircraft. However, the use of intertial navigation in the later part of the program has done much to resolve ambiguities.

Robin's map showed the extent to which glaciers and ice streams persist as thicker lobes of ice within the ice shelf. The map was used to provide preliminary extimates of the directions and speeds of ice flow before more direct measurements became available.

A more detailed map resulted from RIGGS. Navigational problems were minimized by tying the flight lines directly to the network of surface stations whose positions were accurately determined. The resulting map (Figure 73), presented by Bentley, et al. (1979) [37], shows even more clearly the general complexity of the ice thickness map. This complicated pattern of ice thickness presumably reflects complicated dynamics of the ice shelf, and is not associated solely with points of grounding such as Rossevelt Island and Crary ice rise, although large disturbances are found there. The lobes of thicker ice associated with the glaciers and ice streams are shown very clearly. The largest ice stream lobes are those at the top of the map and grid north and east of Roosevelt Island, coming from ice streams B and D, respectively. The most prominent glacier lobes shown are those associated with Nimrod and (especially) Byrd Glaciers; lobes associated with Beardmore and Darwin Glaciers also occur, as can be seen on Robin's (1975) map. The RIGGS map is detailed enough to show a number of enclosed maxima and minima in ice thickness, at least one long linear feature extending along a flow line nearly across the entire ice shelf from Beardmore Glacier in a region of converging ice flow, and several very abrupt changes in ice thickness.

NORTH AMERICA

Radar mapping of a number of glaciers and ice caps in Arctic Canada has been carried out. The first measurements were by air over northern Ellesmere Island in 1966 (Hattersley-Smith, 1969 [101]; Hattersley-Smith et al., 1969 [102]). Measurements over five ice caps, 10 glaciers and the Ward Hunt ice shelf were carried out. In the ice caps the maximum depth measured was 900 m; very irregular topography was indicated beneath the major ice caps, whereas ice depths ranged from 100–250 m for the minor ice caps. Ice depths under the major glaciers ranged from 600–700 m. On the north coast of the island the fjords were shown to have sub-

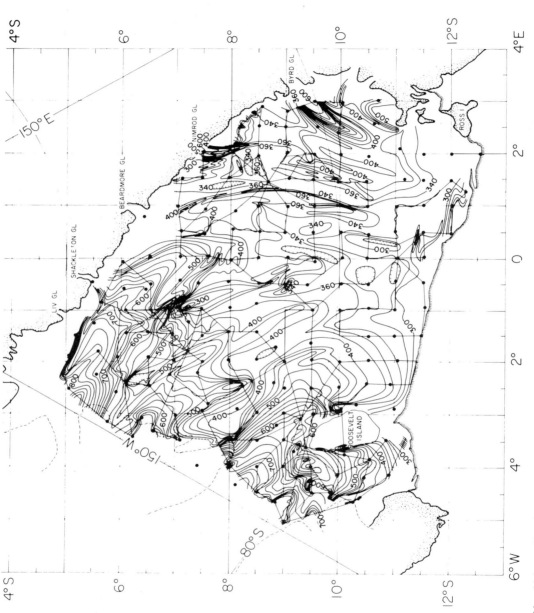

Fig. 73(a). Map of ice thickness, Ross ice shelf, from Ross Ice Shelf Geophysical and Glaciological Survey (RIGGS) data. Contour interval is 20 m. (From Bentley *et al.*, 1979.)

glacial extensions far to the south of the ice fronts of feeding glaciers. A subglacial ridge was discovered beneath part of Otto Glacier, a glacier that is known to have surged. Unexpectedly great ice thicknesses of 70–80 m were found on Ward Hunt ice shelf, and ice rises in the ice shelf were found to be grounded below sea level over their whole extent.

Profiles across Penny ice cap, on Baffin Island, presented by Weber and Andrieux (1970) [187] (Figure 74(a)), and on Barnes and Meighen ice caps, also on Baffin Island (Jones, 1973 [113]), showed the mountainous subglacial topography typical of the region. In 1970, sounding was carried out on four more Arctic ice caps, on Devon, Ellesmere, Meighen and Melville Islands. Maps of the first three ice caps are presented by Paterson and Koerner (1974) [142] – the Melville ice cap was too thin for effective sounding. The mapping on Devon and Ellesmere Islands was confined to detailed studies of small areas of particular interest.

The map of Meighen ice cap differs appreciably from a previous map drawn on the basis of gravity surveys. In the central part of the ice cap the radar soundings gave greater values than the gravity method, whereas the reverse was true in the southern part. On Devon Island soundings were carried out on a small area of about 80 km^2 on the crest of the ice cap. The contour map shows a picture of flat topped hills and deep valleys with ice thicknesses from about 300 m to more than 750 m. All the bedrock in this area is higher in any of the ice-free parts of Devon Island. The small area that was surveyed on Ellesmere Island was on the highest

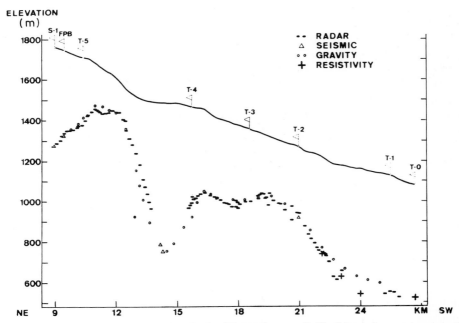

Fig. 74(a). Profile of the southwestern flank of Penny ice cap, Baffin Island, from radar and other sounding methods. The deepest observed radar sounding was 500 m (From Weber and Andrieux, 1970.)

part of the ice cap and consisted of an ice divide on the surface where the surface slope is to the southeast and southwest. The bedrock topography is well dissected and dominated by an eastward trending valley for which there is very little surface expression. The second area sounded is part of the upper catchment area of Otto Glacier. Here, a thickness of 820 m was measured, disclosing a bedrock ridge neighboring a deeply dissected valley, both of which show pronounced surface expression.

A radar sounding system specifically developed for work on temperate glaciers (Goodman and Terroux, 1973 [86]; Goodman, 1975 [84]; see Chapter 4) was first tried out on Athabaska Glacier and Wapta Ice Field in the Canadian Rockies. A contour map of a small part of Wapta Ice Field and a profile across Athabasca Glacier are presented by Goodman (1975) [84]. The same system was used for more complete studies on Rusty and Trapridge Glaciers, yielding quite detailed maps (Clarke and Goodman, 1975 [48]; Goodman *et al.*, 1975 [85]). Cross-profiles of Trapridge Glacier showed a maximum discrepancy where two profiles intersected of 10 m. The maximum measured ice thickness was 143 m in the accumulation region. A small distance downstream there is a bedrock high corresponding to a crevasse field, the maximum observed flow velocity, and a rapid change in surface slope. These ice thickness measurements are of great importance in studying the causes of the surges of Trapridge Glacier.

In 1976 further soundings were made on Devon and Ellesmere Islands and soundings on Axel Heiberg Island were added (Koerner, 1977) [115], all using the 620 MHz radar system developed by Goodman (1975) [84]. Ice depths were found to be generally between 300 and 800 m. The bedrock topography is everywhere very irregular. There is a pronounced difference between the thickness of ice on the east and west sides of the Ellesmere and Devon ice caps. The greater thicknesses on the east sides probably result from higher snow accumulation rates there.

The University of British Columbia monopulse radar has been used for glacier soundings in the Canadian Rockies. The measured ice thicknesses in the Columbia Icefield, $52°10'N$, $117°20'W$, were 100–365 m with an uncertainty of about 10 m; the maximum was much less than the 600–900 m that had previously been estimated (Waddington and Jones, 1978) [177]. The longitudinal profile of the lower part of Wedgemount Glacier, $50°09'N$, $122°48'W$, shows a 'giant staircrase' typical of glaciated valleys, with a maximum thickness of 150 m at the base of the large step (Tupper *et al.*, 1978) [172].

GREENLAND

Figure 74(b) shows the subglacial relief of Greenland. Greenland is covered by a major Arctic ice sheet. Its subglacial relief was investigated by Danish, American, and British scientists using radar sounding. The bedrock topography of Greenland is not homogeneous. In the meridional direction the island is intersected by a trough, like a wide corridor, bounded by highlands that in the south, east, and west merge with high mountain ranges. The maximum height of the eastern range is 1800 m.

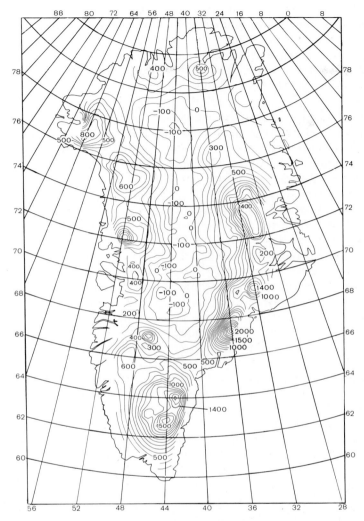

Fig. 74(b). The subglacial relief of Greenland.

SOVIET MEASUREMENTS

Ice bottom topography, with a few exceptions, is identical with that of the underlying rock. Exceptions are those areas where the ice bottom rests on subglacial lakes. Measurements made in Antarctica indicate that the area of subglacial lakes is not large. Hence, studies of the spatial variations of bottom topography would mean the study of bedrock topography. These measurements were started in the 10th SAE. Radar sounding data are extensively used as the basis for detailed maps of bottom relief, which has been explored over large areas.

Radar surveys of the bottom relief are quite detailed (soundings are made every kilometer). This high resolution facilitates the analysis of certain bottom relief features that are not indicated on maps made on the basis of radio-barometric surveys.

Roughness and irregularity of the surface are the most characteristic and marked features of the subglacial topography of the surveyed part of Antarctica, with frequent troughs and peaks, their width being several times larger than their vertical dimensions. This is inconsistent with the concepts on which the available map of Antarctic ice bottom topography in *Atlas of Antarctica*, 1966 [1] was constructed.

Radio barometric surveys have revealed a subglacial strait extending from Edward VIII Bay to Amundsen Bay, a larger mountain range south of the strait (between 68° and 69°S), a sub-glacial plain bounded by mountains in the vicinity of Showa Station, and another, larger, mountain range south of Showa Station, running from 72°S southward.

In the 17th SAE (1972), a detailed survey of the subglacial topography was made during an inland tractor traverse. Two polygons were established for this purpose along the track from Mirny to the 153rd km towards Pionerskaya. The polygons, measuring 5 by 5 km, were emplaced at the 57th and 153rd km of the track. Radar sounding was made along criss-crossing lines 500 m apart at each polygon, lines being mutually perpendicular. These measurements were used to make detailed two dimensional maps of subglacial topography, thus helping to relate the stake-bounded areas of the polygons to the bottom features. Figure 75(a) and 75(b) illustrate the usefulness of radar techniques in the study of two-dimensional structure of ice bottom. It is seen in the left-upper corner of Figure 75(b) that no reflected signals were recorded. This suggests a steep slope.

The Soviet Arctic is the third-largest glaciated province of the Earth (after Antarctica and Greenland). Large ice masses cover the islands of Novaya Zemlya,

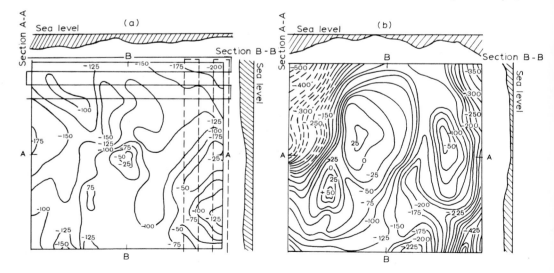

Fig 75. Subglacial topography on polygons: a – 57 km from Mirny Observatory; b – 153 km from Mirny Observatory. The lines at the top and in the right-hand part of figure (a) are sample radar-sounding tracks.

Severnaya Zemlya and Franz-Josef Land). The first radar soundings of arctic glacier thicknesses were started in 1968. Since 1975 they have been carried out systematically – Figure 76 is a map of Severnaya Zemlya glaciers. Glaciers in Svalbard have been explored by the Geographic Institute of the U.S.S.R. Academy of Sciences. The Arctic glaciers are known to be of relatively small dimensions and thicknesses (less than 1000 m). There are a number of reference points in their periglacial zones that providing for precise closure of the sounding tracks. Local glaciological and meteorological conditions in the Arctic determine the time when airborne radar measurements are possible: no meltwater on the surface, no strong katabatic winds, etc. Hence March, April and early May are the most convenient months for radar surveys. Figure 77 presents several characteristic profiles of glaciers and island ice domes. Their ice topography and subglacial rock morphology are reconstructed by radar measurements taken along the lines shown in Figure 76.

A radar survey of the thickness of Vavilov Glacier was carried out in 1977–79. It resulted in a map of the subglacial topography (Figure 78). The thickness of the Severnaya Zemlya glaciers is shown in tabular form by Bogorodsky et al. (1970) [12]. Macheret and others (1980) [17], (1973) [18] discuss radar measurements on

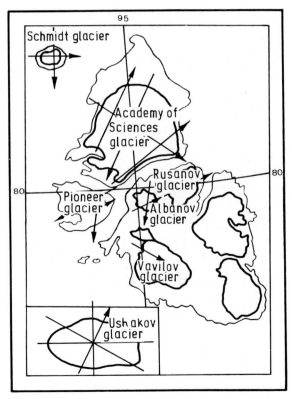

Fig. 76. Glaciers of the Severnaya Zemlya archipelago. Lines correspond to the routes of airborne radar sounding.

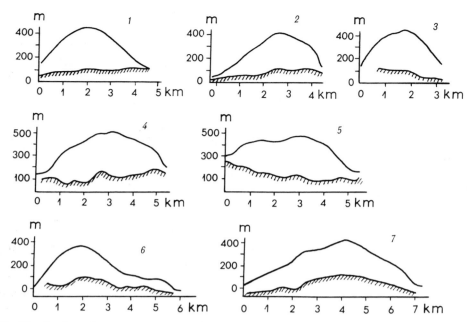

Fig. 77. Sections of Severnaya Zemlya glaciers from airborne radar sounding. Ice thickness in meters is shown along the vertical axis. 1, 2, 3 – three sections across Pioner Dome; 4, 5 – two sections across Albanov Dome; 6, 7 – two sections across Schmidt Dome.

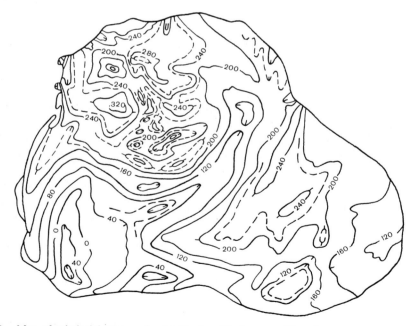

Fig. 78. Map of subglacial topography for Vavilov Glacier area. Elevations on contour lines are in meters.

the Svalbard glaciers carried out by the Geographic Institute of the U.S.S.R. Academy of Sciences.

The foregoing examples indicate the promising aspects of radar sounding techniques for the study of the morphology of subglacial geological structures.

6.7. Subglacial Physiography and Geology

Studies of the characteristics of the subglacial bed can yield a substantial amount of information about the subglacial geology. Such information becomes increasingly useful as the distance to the nearest outcrop increases — consequently, most of the studies of this kind have been made in Antarctica.

ANTARCTICA

A start on such studies was made by Beitzel (1971) [32] who compared the subglacial physiography of Queen Maud Land with the Basin and Range province of the western United States. The leader in this type of work, however, has been D.J. Drewry. In his first paper on the subject, Drewry (1971) [66] examined the subglacial topography between the Southern Transantarctic Mountains and the South Pole. In this region the radar-sounding results indicate a generally uniform transition between the mountainous margin and the continental interior and show no evidence of major structural discontinuities. Where there is sufficient data control, the subglacial terrain reveals a pattern of ridges and troughs, some of which trend towards the plateau, indicating earlier drainage lines. These troughs, together with those penetrating the Transantarctic Mountains, probably stem from earlier local glaciation in which active, more temperate, ice streamed down both sides of the mountains. With the onset of full-scale continental glaciation, the structurally lower inland region became submerged by the progressive build-up of ice in the interior, reversing the drainage on that side with little modification of the earlier topography.

Continuing his study northward towards Victoria Land, along the Transantarctic Mountains, Drewry (1972) [67] showed that the radar studies confirm surface geological investigations in indicating the complex pattern of differentially tilted fault blocks. Some of these blocks extend as far as 600 km from the Ross Sea coast into Victoria Land. There are variations in the magnitude of tilting and degree of secondary longitudinal and transverse faulting within the blocks. These factors, combined with varying amounts of erosion, have produced a complicated transitional zone between the epi-cratonic mountain belt and the lowland shield of the East Antarctic craton.

Analyzing the radar profiles from the Wilkes Land area by means of the autocorrelation and standard deviation of the relief, Drewry (1973, 1975) [68, 70] was able to differentiate between topographic areas in East Antarctica in terms of the degree and scale of roughness displayed by the subglacial terrain. A pilot study of terrain anisotropy was also useful in that regard.

Specifically, Drewry (1975) [70] found that three subglacial topographic groups could be identified (Figure 79), one comprising lowland areas and the other two

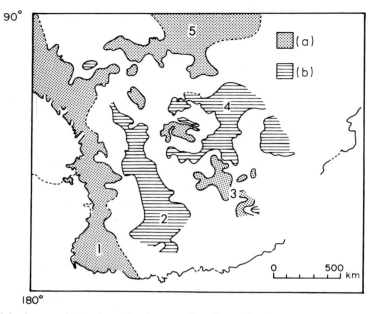

Fig. 79. Major large-scale terrains units of eastern East Antarctica. 1 – Transantarctic Mountains; 2 – Wilkes subglacial basin; 3 – unnamed highland massifs; 4 – Aurora subglacial basin; 5 – Gamburtsev subglacial Mountains. Stippled areas, a, subglacial elevation >250 m; lined areas, b subglacial elevation <−500 m. (From Drewry, 1975.)

highland regions, one with a relatively smooth and regular topographic surface but dissected by deep valleys, and the other showing a very rugged mountainous terrain. Highland terrains of both types are found within the Gamburtsev Subglacial Mountains. Such contrasts within continuous mountain zones probably originate from differences in the geological structure. If the Gamburtsev Mountains, as has been suggested, were formed from a thick geosynclinal rock sequence around the end of the Precambrian, then the presence of two distinct provinces suggests more than one episode of block formation and uplift. It may be that periods of reactivation in the Gamburtsev Mountains did not have a uniform effect over time or space, leading to later diversification of terrain. On the other hand, the differences might be the product of differential erosion of a sedimentary cover occurring during the Phanerozoic.

Within the lowland areas, Drewry (1976) [71] has used the radar-echo results along with other geophysical information to deduce the presence of two major sedimentary basins: the Wilkes Basin to the east and the Aurora Basin to the west. Sediment appears to be extremely thin or absent on the inland flank of the Wilkes Basin where terrain roughness becomes significantly greater than along the axial core of the basin. Furthermore, radar measurements indicate sudden breaks in slope which point to possible fault control and deep erosion of the basin's edges. The thickness of sediments filling the Wilkes Basin is probably less than 3 km, because bedrock highs penetrate the cover in some localities. Much less information is available for interpretation of the Aurora Basin. Sediment thicknesses are unknown, but a pattern of smooth basins and intervening rough basement highs is suggested from radar and other geophysical work.

The Transantarctic Mountains and massifs within central Wilkes Land show statistical similarities, but both are quite different from the Gamburtsev Mountains (Drewry, 1975) [70]. The structural control of relief in both the exposed and the subglacial parts of the Transantarctic Mountains is partially attributable to the presence of the subhorizontal sediments of the Beacon Supergroup and associated intrusive bodies. Values of terrain roughness indicate that the central massif of Wilkes Land also falls within the same province of Phanerozoic sedimentation. Drewry (1973) [68], in addition, applied a model taking account of the stiffness and flexural properties of the crust to determine the isostatically-compensated preglacial topography of eastern East Antarctica, and also used empirical relationships between bedrock and depth to the Mohorovicic discontinuity to map the elevation of the base of the crust.

Drewry (1973) [68] further suggested that the profiles and relief maps make it possible to delimit regional neo-tectonic zones of depression and uplift, including deep trenches which may be rift-like in structure. The combination of neo-tectonic studies, terrain statistics, and other geophysical measurements led him to a generalized outline of the tectonic pattern of central East Antarctica that is somewhat different from that of previous Soviet investigators, including changes to the inland boundary of the Ross orogenic belt and the identification of two new tectonic provinces.

In a detailed study of the interior of Wilkes Land, Steed and Drewry (1982) [165] note that many valleys and re-entrants within the subglacial massifs show glacial characteristics. It is possible that they date from a temperate phase of upland glaciation prior to the establishment of the ice sheet.

Even more detailed radar, magnetic, gravity, and seismic measurements have been made on the ice sheet surface in the vicinity of Dome C within Wilkes Land. This combination is powerful one for studying subglacial geology. Topographic variations of subglacial rock exert great effects on gravity values, from which it is possible to estimate the density of the underlying material. On the other hand, the magnetic anomalies may or may not reflect the bottom topography, leading to a clear conclusion as to whether the immediately-subglacial rocks are magnetic or not. The analysis at Dome C (Blankenship et al., 1980 [40]; Gassett et al., 1980 [81]) showed that the rock forming the subglacial terrain has a density close to 2.7 mg m^{-3}, and that the high relief in the area must be caused by variations in rock of high magnetism. Combined with seismic refraction studies, these data indicate a highly magnetic crystalline basement immediately beneath the ice. This shows that, in this area at least, there is no sedimentary layer of significant thickness between the ice and the basement.

The edge of the ice sheet near McMurdo Sound has received a comprehensive analysis by Calkin (1971 [45]; 1974 [46]), using the results of airborne radar sounding over Wilson Piedmont Glacier, Mackay, Ferrar and Taylor outlet glaciers, glaciers in Victoria and Wright Valleys, and the ice sheet bordering the mountains. Calkin (1971) [45] concludes that Victoria and Wright Valleys have different geomorphic histories. A smooth subglacial profile of lower Wright Valley is more compatible with repeated outlet glaciation than is the rougher profile of the lower Victoria Valley, where there is evidence of an important amount of erosion by local glaciers. Calkin (1971) [45] also found evidence for marginal faulting along the edge of McMurdo Sound.

Calkin (1974) [46] further finds that the floors of the Debenham, Wright, and Victoria Valleys occur beneath the Wilson Piedmont Glacier at elevations of about 260 m, 260 m, and 670 m, respectively. The 670 m threshold may have blocked easterly marine and glacial invasions experienced by lower valleys. Profiles along the outlet glaciers display large depressions associated with erosion by tributaries and with glacial erosions through thick dolerite sills. The subglacial west flank of the mountains is formed by a series of high, steep-sided plateaus with gentle westward-sloping surfaces. Block faulting, westward-dipping dolerite and sandstone units, and glacial erosion must explain this topography. Alpine glaciers and westward movement of the incipient ice sheet of East Antarctica must also have contributed to the overall subglacial topography of the west flank of the mountains here. Inland-sloping valleys heading above Mackay and Mulock outlet glaciers may be, at least partly, of glacial origin.

Systematic airborne radar sounding was conducted along lines only 10–15 km apart inland of the dry valley region in 1974–75 (Drewry, 1982) [72]. The subglacial topographic map shows the continuation of mountainous topography declining in elevation westwards. Combining the results of seismic refraction profiles with reflection and roughness characteristics from radar sounding indicates the presence of Beacon-type terrain beneath the ice sheet, although deep valleys appear to expose basement rocks.

Rose (in press) has analyzed radar studies of bedrock characteristics in western West Antarctica. The bedrock is generally below sea level, being smooth near the ice shelf and becoming rougher towards the east. The smooth area bordering the Ross ice shelf is believed to be part of a large sedimentary besin which extends into the Ross Sea. Within this basin there are several prominent bedrock plateaus possessing very smooth surfaces; these are interpreted by Rose as having been leveled off by the base of a partly floating ice sheet.

The more easterly part of West Antarctica, south of the Antarctic Peninsula, has been studied by Drewry *et al.* (1980) [74] and Jankowski and Drewry (1981) [106]. In this area, radar sounding was carried out simultaneously with an aeromagnetic survey, which improved the geological interpretation. A pronounced topographic boundary exists between the Byrd Basin and the rugged Ellsworth Mountain block, also marked by an abrupt change in the magnetic pattern. Together, these data suggest that a geological sequence similar to that exposed in the Ellsworth Mountains extends to the westward edge of the subglacial block. Magnetic anomalies associated with a sinuous ridge running across the Byrd Subglacial Basin, delineated by radar sounding, suggest that the feature may be of volcanic origin, and further lends weight to the suggestion that the Byrd Subglacial Basin may contain a sequence of interbedded sedimentary and volcanic strata overlying the basement. The radar data also show that the transition from West to East Antarctica is abrupt. On the East Antarctic side the bedrock terrain is extremely rugged and dissected, and lies close to the ice sheet's surface in the vicinity of the exposed mountains – a zone more than 150 km wide. This subglacial surface gradually becomes smoother and deeper towards the interior of East Antarctica. There is an abrupt change on the West Antarctic side, however, where the bedrock lies approximately 1500 m below the ice sheet's surface.

Wyeth (1977) [194] has used airborne radar data from Smith (1972) [163] to study the transition zone between Graham Land and Palmer Land in the Antarctic Peninsula. The rock surfaces beneath the snow plateaus forming the spines of these two parts of the peninsula differ considerably in character. The ice is much thicker in Palmer Land and the subglacial topogrpahy is very uneven with slopes rising more than 1000 m over a horizontal distance of 2 km. In contrast, the spine of Graham Land is a fairly flat rock plateau. These data have been used, together with other geological and geophysical evidence, to support the interpretation that an important tectonic break in the Antarctic Peninsula occurs between Graham Land and Palmer Land and that the transition zone, at least in its present form, is geologically young.

Drewry (1982) [72] has used the identification of water at the bed of the ice sheet – from radar sounding (Section 6.10) – as an aid in estimating geothermal flux. Recognition of water at the bed provides a limit to the temperature profile through the ice sheet. This observation permits calculation of the basal temperature gradient, and therefore the basal heat flux from below. Using a simple steady-state one-dimensional heat flow equation (thus assuming that horizontal advection of ice is negligible near the center of an ice sheet), and picking reasonable values for the mean values of the temperature and snow accumulation at the surface, Drewry calculated a temperature gradient beneath the ice just inland of the dry valley region of $50°C\,km^{-1}$, and a heat flow of 2.0 to 2.7 HFU (0.08 to 0.11 $W\,m^{-2}$). These values are significantly higher than the global average, but are consistent with measurements in the dry valleys.

GREENLAND AND NORTH AMERICA

Robin et al. (1969) [153] commented briefly on the correspondence between the roughness characteristics of the bedrock beneath the ice sheet in north-west Greenland, and the observed characteristics of the mountainous coastal strip, finding the two compatible. In their more detailed study of Roslyn Gletscher in East Greenland, Davis et al. (1973) [62] were able to define a subglacial ridge which continued downstream from the confluence of the main glacier and a large tributary, and suggested that the two streams of ice maintain their own identity far down the valley. They also found evidence of discontinuities within the ice, which they suggest may have resulted from avalanches off the rock walls above the glacier.

Analysis of the radar soundings over Otto Glacier in Northern Ellesmere Island (Hattersley-Smith, 1969) [101] showed that the surge of the glacier in the 1950s allowed the deeper ice stream of the main glacier to push aside the stream from a tributary glacier. Minor subglacial ridges found near the snouts of several glaciers on the south side of the central ice cap on Ellesmere Island may well represent moraines overridden by recent glacial advances, since the glaciers are believed to be at their most advanced position during at least the last 1000 years.

On the Devon Island ice cap, a study of the reflection strength by Oswald (1975) [134] led to the identification of a likely geological boundary, with a change from higher to lower permittivity when traveling from west to east. Unfortunately, it was not possible to attempt identification of the rock types owing to uncertainty in the absolute value of the reflection coefficient, an uncertainty that it should be possible to decrease in future work.

6.8. Radar Sounding of Internal Layering

Thus far we have been concerned principally with reflections from the base of the ice. The radar technique can also be used to detect discontinuities in the ice that give significant changes in the dielectric parameters over distances small compared with the pulse length. The presence of water in an ice or snow mass can lead to large effects because of the very strong dielectric contrast. However, internal reflections are by no means limited to melt zones. In fact, internal layering is a striking feature of radar reflection surveys over cold glaciers and ice sheets, particularly Antarctica and Greenland. The nature of the internal reflectors is a matter of some dispute. Reflections have been attributed to such causes as microparticle concentrations (e.g., volcanic ash layers), density contrasts (due either to melting at the surface creating icy layers or lenses, or to strain-induced variations in bubble content), crystal anisotropy, and variations in chemical impurities which would affect primarily the imaginary, rather than the real, part of the dielectric constant. Most of the theories relating to the cause of the layering attribute it to some phenomenon that takes place at the surface of the ice sheet; consequently the implication is strong that these layers can be taken as isochrons, particularly at substantial heights above the base of the ice. That fact makes the layers very useful in plotting flow lines within the ice, and in looking for indications of variations in those flow patterns with time.

Internal layering is not the only kind of internal reflection found by radar surveys on glaciers and ice sheets. Other sources of reflected energy include brine infiltration layers, bottom crevasses and buried former surface crevasses, morainal layers entrained in the ice, and water inclusions in temperate glaciers.

THEORY OF DETECTION OF INTERNAL LAYERS

From an electromagnetic point of view the explanation of the source of internal echoes is simple: any change in permittivity will give rise to a reflection of incident electromagnetic energy, and in the present case it is obvious that, for one reason or another, the snow deposited on the ice sheet from time to time has had characteristics different from normal, and thus has created a more-or-less horizontal stratification of ice, each layer deviating in permittivity from the ice below and above. This stratification will be dealt with in another section, but two questions arise in the context of an analysis of the measurement procedures: what is the possibility of detection of layers with a certain sounding system, and what is the permittivity deviation necessary to create a return signal which may be detected by the system?

If we assume that the reflecting horizons in the ice sheet are caused by thin horizontal layers in which the permittivity deviates from the surrounding ice permittivity, we may calculate the reflection loss for changes in the real and the imaginary parts of the permittivity separately. With the complex permittivity given by

$$\varepsilon = \varepsilon_0 \varepsilon' (1 - i \tan \delta), \tag{6.7}$$

we find that

$$\Delta \varepsilon = \varepsilon_0 \Delta \varepsilon' - i \varepsilon_0 \varepsilon' \Delta \tan \delta, \tag{6.8}$$

or the relative variations

$$\frac{\Delta \varepsilon}{\varepsilon_0 \varepsilon'} = \frac{\Delta \varepsilon'}{\varepsilon'} - i \Delta \tan \delta, \tag{6.9}$$

and we may operate with the first and the second term as cases of variation.

For a layer thickness of h and a relative permittivity change of $\Delta \varepsilon / \varepsilon$, we find a reflection coefficient

$$R = \frac{i \tan \theta (1 - W^2)}{2W + i \tan \theta (1 + W^2)}, \tag{6.10}$$

where $\theta = Wh = 2\pi h/\lambda$ is the electrical length of the layer and $W = \sqrt{\varepsilon_2/\varepsilon_1}$ is the ratio between the wave impedances in the surrounding ice and the layer, respectively. Inserting $W = \sqrt{\varepsilon_2/\varepsilon_1} = \sqrt{1 + (\Delta \varepsilon/\varepsilon_1)}$, we find the power reflection coefficient

$$|R|^2 = \frac{\tan^2 \theta (\Delta \varepsilon/\varepsilon_1)^2}{4\left(1 + \frac{\Delta \varepsilon}{\varepsilon_1}\right) + \tan^2 \theta \left(2 + \frac{\Delta \varepsilon}{\varepsilon_1}\right)^2}$$

$$\simeq \frac{\theta^2 \left(\frac{\Delta \varepsilon}{\varepsilon_1}\right)^2}{4(1 + \theta^2)} \simeq \left(\frac{\pi h}{\lambda}\right)^2 \left(\frac{\Delta \varepsilon}{\varepsilon_1}\right)^2, \tag{6.11}$$

where we have assumed that θ is small and $\Delta \varepsilon / \varepsilon_1 \ll 1$. Taking $\theta < 6°$, for instance, we find that $h/\lambda < 0.0167$, or at 60 MHz with $\lambda = 2.81$ m, $h < 4.6$ cm. Assuming only $\Delta \varepsilon / \varepsilon_1 \ll 1$, (6.11) becomes

$$|R|^2 = \frac{\tan^2 \theta (\Delta \varepsilon/\varepsilon_1)^2}{4(1 + \tan^2 \theta)} = \frac{1}{4} \left(\frac{\Delta \varepsilon}{\varepsilon_1}\right)^2 \sin^2 \theta. \tag{6.12}$$

Based on this equation, Figure 80 has been produced, showing the reflection loss, $20 \log |R|$, as a function of the relative layer thickness h/λ with $\Delta \varepsilon/\varepsilon_1$, as a parameter. It is seen that even thin layers with small deviations give appreciable reflections. Thus, a reflection loss of 70 dB is obtained for a layer with a relative thickness of 0.05 (14 cm at 60 MHz) and $\Delta \varepsilon/\varepsilon_1 = 0.002$ ($\Delta \varepsilon = 0.006$); the same reflection loss being obtained for a relative thickness of 0.01 (3 cm at 60 MHz) and a $\Delta \varepsilon/\varepsilon_1 = 0.01$ ($\Delta \varepsilon = 0.03$).

From (6.9) it is clear that we may also use Figure 80 for the case of changes in the imaginary part of the permittivity by replacing $\Delta \varepsilon/\varepsilon_1$ with $\Delta \tan \delta$, i.e., the abso-

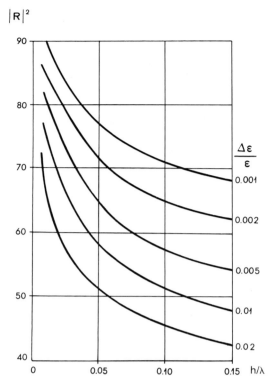

Fig. 80. Reflection loss, $|R|^2$, as a function of h/λ for ice layers with different values of $\Delta\varepsilon/\varepsilon$.

lute change in the loss tangent. It is seen that small changes (which may difficult to verify by laboratory measurements on ice samples) will give detectable reflections. In fact, we may use the calculations of the excess loss, by means of (4.15), to work out estimates of the detectability of layers at a certain depth in the ice and with a certain reflection loss if we assume the dielectric loss in the ice above the layer. From the measured and the calculated temperature depth profiles in an ice sheet, it is known that the temperature is approximately constant in the upper layer of the ice cap, i.e., from a depth of 10–20 m down to a depth of about 500 m above the base. In the uppermost part (in the firn), the temperature varies with the season but, since the ice density has a low value, the contribution to the total loss becomes small and the effect may be disregarded in our approximative calculations. In contrast, in the lower part of the ice cap the temperature increases rather drastically, so that a good estimate may be made only if the temperature profile is known with some accuracy.

Thus, concentrating alone on the upper part of the ice sheet, we may assume a constant temperature, and thereby a constant loss, down to the level where basal heating sets in. Using the previously-derived diagram for the excess loss $L_A L_R$ (Figure 21) as it results for a certain 60 MHz system (pulse length 250 ns and bandwidth of 4 MHz) and the laboratory data for dielectric loss versus temperature

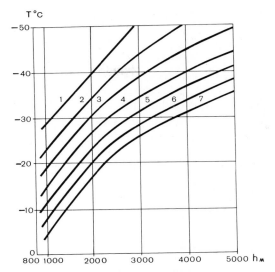

Fig. 81. Reflection loss for a layer at the depth shown on the abscissa, as a function of the temperature, shown on the ordinate. $f_p = 60\,\mathrm{MHz}$, $\tau_t = 250\,\mathrm{ns}$, $\Delta f = 4\,\mathrm{MHz}$; N_R: 1 – 120; 2 – 110; 3 – 100; 4 – 90; 5 – 80; 6 – 70 and 7 – 60 dB.

(exponential approximation $0.0588\exp(T/15.9)$, we arrive at Figure 81. This diagram shows the depth of a reflecting layer (abscissa) with a certain reflection loss (parameter), which may just be detected with a system (whose parameters are known) through a column of ice at a constant temperature, as indicated by the ordinate. It is seen that a layer with a reflection loss of 90 dB may be detected by the system in question at a depth of 2400 m if the temperature in the ice column above the layer is lower than −30°C. This example is relevant for sounding in central Greenland, where the temperature of the upper ice is about −31°C and the total ice thickness may attain 3200 m. In East Antarctica, where the ice temperature over large areas is about −50°C and the total thickness is about 4500 m, we may detect layers with reflection loss up to 100 dB to depths where the basal heating becomes of importance, perhaps 3500 m.

From Figure 80 it is seen that the layers mentioned above may be very thin and have very small changes from the surrounding ice. To obtain a reflection loss of 90 dB we may assume a relative thickness of the layer of 0.01 ($h = 3\,\mathrm{cm}$ at 60 MHz) and $\Delta\varepsilon/\varepsilon_1 = 0.001$, for instance. Clearly, the system used here as an example is a very powerful one that is likely to detect stratification of almost every magnitude to great depths in cold polar ice.

From the formula for excess noise (4.15) it is seen that Figure 81 may also be used for systems other than the one used as an example if their frequencies are in the range 10–300 MHz where the ice losses are independent of frequency. Thus, any reduction or improvement in system sensitivity (expressed in decibels) will be reflected in the magnitude of L. In the example system, the pulse length and the corresponding bandwidth were chosen to be $\tau_t = 250\,\mathrm{ns}$ and $B = 4\,\mathrm{MHz}$, respectively. If instead we had chosen $\tau_t = 60\,\mathrm{ns}$ ($B = 14\,\mathrm{MHz}$) the system sensitivity would have been reduced by about 5.5 dB, so that the reflection loss determined by Figure 81 should be reduced by the same magnitude. In case Figure 81 is used for a frequency different from 60 MHz, it should be recalled that the parameter c_2 in (4.16) contains a wavelength dependence (squared). Therefore, for a system with the same characteristics as those in the example system except for the frequency, L_R should be adjusted by the value $20 \log (60/f_p)$. For a 300 MHz system the reflection loss becomes 14 dB smaller than obtained from Figure 81, for instance, due to the reduction in system sensitivity with the increase in frequency. Figure 81 may be taken only as an estimate of the possibility of detection of a layer at a certain depth with a given reflection loss. For a more accurate estimate a temperature profile measured at the point in question should be used. One of the few places where a profile was measured is at Camp Century in northwest Greenland; we shall use this as an example of the results obtainable. As a simplification, we will divide the 1387 m (1400 m) into 14 100-meter layers, for which we employ the average temperature as given in Table XVIII. It is seen how the basal heating sets in at about 1000 m (with the base temperature measured at $-13.5°C$). The table also gives values for L_R calculated as shown previously for the 60 MHz system used as an example. It is seen that the basal heating only reduces the detection capabilities by about 4 dB. The average temperature for the first 1000 m ice is about $-23°C$, so that L_R becomes 111.4 dB according to Figure 81, whereas Table XVIII gives a value of 111.7 dB. One may conclude that Figure 3 is a rather good estimate.

TABLE XVIII

Detection of layers by radar sounding by means of the 60 MHz system with $\tau_t = 250$ ns and $B = 4$ MHz

Depth (m)	Temperature (°C)	Reflection loss (dB)
100	−24.4	131.6
200	−24.7	126.6
300	−24.6	123.3
400	−24.3	120.7
500	−24.0	118.6
600	−23.6	116.8
700	−23.1	115.3
800	−22.3	113.9
900	−21.3	112.7
1000	−20.1	111.7
1100	−18.5	110.7
1200	−17.0	109.8
1300	−15.5	109.0
1400	−14.0	108.3

The conclusion from Figure 81 is that a reduction of the system's sensitivity does not give very great reduction in the depth at which a layer with a certain reflection loss may be detected. Let us, as an example, consider a case where the ice has a temperature of $-30°C$. From Figure 81 we see that a 10 dB reduction in sensitivity reduces the depth from 3840 m to 3370 m for a reflection loss of 60 dB, for instance, or 13%. For a 80 dB reflection loss the corresponding number is 21%. Although relatively small, the difference may be crucial if the layers have a reflection loss of 80 dB or higher. Some of the early systems had a system sensitivity of 160 dB compared with the one used here as an example which has a sensitivity of 184 dB (for same pulse length of 250 ns, $B = 4$ MHz). From Figure 81 it is clear that this 24-dB reduction in performance will reduce the possibility of detecting layers if their reflection loss is about 80 dB or higher, except in very cold ice.

INTERNAL LAYERING

As previously mentioned, internal echoes have been a striking characteristic of all radar sounding carried out on the Antarctic inland ice sheet. Jiracek (1967) [110] found strong internal reflectors not only at South Pole Station and on Roosevelt Island, but also on both the grounded and floating parts of Skelton Glacier. Harrison (1973) [100] concluded from a theoretical analysis that the internal reflectors in the Antarctic ice sheet (and Greenland as well) result from sequences of thin layers rather than single discrete discontinuities. This analysis led him to favor crystal anisotropy as a major, if not the only, cause of the dielectric variations. Clough (1977) [55] made a study of the internal reflectors at Byrd Station, Antarctica, in comparison with the density variation and microparticle content within the core from the deep drill hole at that station. He showed a good correlation between the two, strongly suggesting that for reflectors down to 1000 m or so density variations are the cause of reflections. Clough also showed that despite the apparent continuity of the internal layers over large distances when viewed broadly, in detail the layers are not continuous, nor are they perfectly flat. This may, in part, be due to insufficient resolution of successive layers (Clough, 1974, [52]). Further recent studies at Dome C show, from detailed continuous recording on the surface, that the internal reflectors are rather more like dashed lines than solid lines.

Studies in both Antarctica and Greenland have almost always shown an absence of internal reflectors in the basal zone of the ice sheets. It is a very important problem to determine whether that absence is due simply to insufficient sensitivity of the equipment or whether there is a true absence of discontinuities within that zone. If the latter is true, then, as pointed out by Robin et al. (1977) [152], the reason is quite likely to be variable shear strain in the lowest layers due to variable deformation over a rough relief. This, in turn, makes questionable the validity of dating ice cores close to bedrock, since the assumptions of uniform deposition and strain have been used in calculating dates at those levels. Caution must be used, therefore, in interpreting the palaeoclimatic record from the lowest layers of the ice sheet or ice cap. Ackley and Keliher (1979) [25] have compared the internal echoes with measured properties of the deep core at Cape Folger in the East Antarctic coastal zone. They believe that the deeper echoes, at least, are due to variations in the bubble structure within the ice.

For reasons that are not clear, internal reflectors of the type we have been discussing seem to be largely absent in the Antarctic ice shelves. Instead, several other types of reflectors are found – we will consider them in the next three sections.

BRINE INFILTRATION ZONES

Whenever sea water gains access to the permeable firn above the firn–ice boundary it can penetrate laterally for distances as great as 10 km. The firn–ice boundary is found at much the same depth over most of the ice shelf, regardless of the ice thickness, so that where the ice is thin the boundary lies below sea level. There are at least two places on the Ross ice shelf where sea water can gain access to such thin ice: the seaward margin of the McMurdo ice shelf south of Ross Island, and along the Transantarctic Mountain front between feeding glaciers, where the ice shelf is thin because the ice input from land is very small. In the latter regions, it appears that cracks through the shelf occur, presumably because of the great stresses occurring where the thin ice along the mountain front adjoins the fast-moving main body of the ice shelf. Sea water can then reach the firn–ice boundary by penetrating upward through the cracks.

Reflections from brine-soaked firn in the McMurdo ice shelf were observed first by Clough and Bentley (1967) [49], although they were recognized at the time only

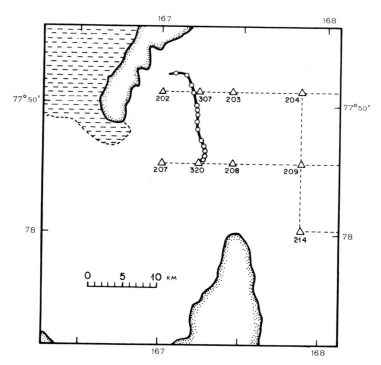

Fig. 82. Map of McMurdo ice shelf showing the location of the edge of the brine infiltration zone (solid line) as delineated by radar crossings (circles). Triangles denote stations of Heine (1968). (From Clough, 1973.)

as a jump in apparent thickness. After finding a similar jump in apparent thickness from 1967 radar flights across the same line, Swithinbank (1970) [168] first made the correct correlation with the brine-soaked layer.

The first detailed study was carried out by Clough (1973) [51] in the course of profiling the McMurdo ice shelf on the grid system partially shown in Figure 82. The abrupt change in apparent thickness occurs about 10 km from the ice shelf's margin. On the western side the depth to the reflector is 32–45 m, compared with 95–120 m on the eastern side. Brine-soaked firn was found at a depth of 19 m in cores taken near Station 207 and at 26 m midway between Station 207 and 320 (Heine, 1968 [103]; also personal communication, Heine to Clough, 1972). The eastern extent of the brine penetration was found by radar profiling about 5.5 km

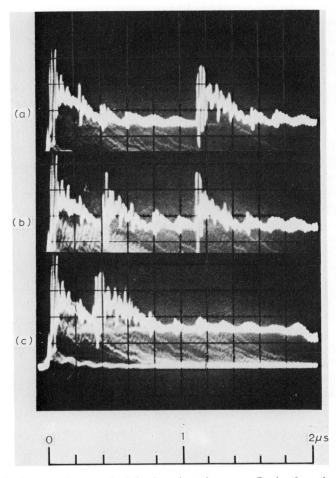

Fig. 83. Reflection records across the brine layer boundary. a – reflection from the bottom of the ice shelf is seen at approximately 1.1 μs; b – reflection from both the bottom of the ice shelf and the top of the brine layer ($\simeq$0.4 μs); c – reflection from the top of the brine layer. The reflection from the bottom of the ice shelf is still visible, but small. (From Clough, 1973.)

east of Station 207 and at an indicated depth of about 32 m. The depth corresponds well with Heine's (1968) [103] estimate of 30–31 m at Station 320.

Following this discovery, a portion of the boundary delineating the extent of brine penetration of the ice shelf was mapped by a zig-zag traverse. The boundary and traverse crossing points are shown in Figure 82. Reflection records obtained in the vicinity of the boundary are shown in Figure 83. The recorded echo changed completely from deep to shallow over distances of a few meters. Moving a few meters further to either side of the boundary normally resulted in extinction of the weak echo seen in Figure 83(a) or 83(c). A very high reflection coefficient at the top of the brine layer and strong absorption within the layer apparently result in no reflection from the bottom of the ice shelf in the region of the brine layer. Right at the boundary it is possible to obtain reflections from both the brine layer and the bottom of the shelf, as can be seen in Figure 83(b). The time shift in the reflection time to the brine layer seen when comparing Figure 83(b) and 83(c) is too great to be caused only by the oblique incidence of the wave; it must be due, at least in part, to the shape of the edge of the brine layer as well.

Using a high resolution impulse radar, Kovacs and Gow (1975 [118]; 1977 [119]) have continued the study of the brine infiltration zone on the McMurdo ice shelf. Their investigations have revealed several featured not previously observed in the records of Clough (1973) [51]. Several of these structures can be seen in Figures 84 and 85. What are believed to be fractures within the ice are shown in Figure 84. These cracks probably afford a route for sea water to rise upward and then spread laterally into the firn. Figure 85 shows the multiple character of the brine layer echo in some places, a characteristic that has not yet been explained.

Reflection from a brine infiltration layer is best known for the McMurdo ice shelf, but is probably also the source of very shallow echoes about at sea level that occasionally occur without any accompanying bottom echo on soundings over the Wordie, Brunt, and Larsen ice shelves and on Wilkins Sound (Smith and Evans, 1972 [164]) (Figure 86). Smith and Evans (1972) [164] have calculated the attenuation to be expected through a layer of brine-soaked firn. Assuming snow of density $0.8 \, \text{mg m}^{-3}$ with pores filled with sea water ($\sigma = 2.9 \, \Omega^{-1} \text{m}^{-1}$) leads to an absorption loss of 38 dB m^{-1}, i.e., a 76 dB loss in a round-trip through a layer 1 m thick. It is thus not surprising that the bottom echo is lost in places where a brine infiltration layer occurs.

Evidence for brine layer occultation of the bottom echo is found extensively in radar reflection records from the Ross ice shelf, particularly over the eastern part of its grid. Several authors have used the distribution of zones of bottom echo occultation to trace flow paths across the shelf (see p. 211ff.).

SCATTERING BY AIR BUBBLES, WATER-FILLED CAVITIES, AND ICE LENSES IN GLACIERS

Robin et al. (1969) [153] considered the propagation of radio waves in clear solid ice containing a distribution of spherical air bubbles. In a polar ice sheet it is appropriate to consider a sparse, random distribution of bubbles; the upper limit to the bubble radius being about 1 mm. In this case a straightforward analysis leads to the

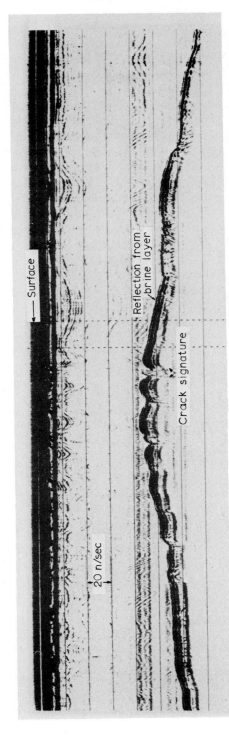

Fig. 84. Impulse radar profile of the brine layer in the McMurdo ice shelf along a section about 2 km from the margin of the ice shelf. Suspected cracks occur along the highest part of the layer, 9–10 m below the surface. Horizontal lines are 20 ns apart. (From Kovacs and Gow, 1975.)

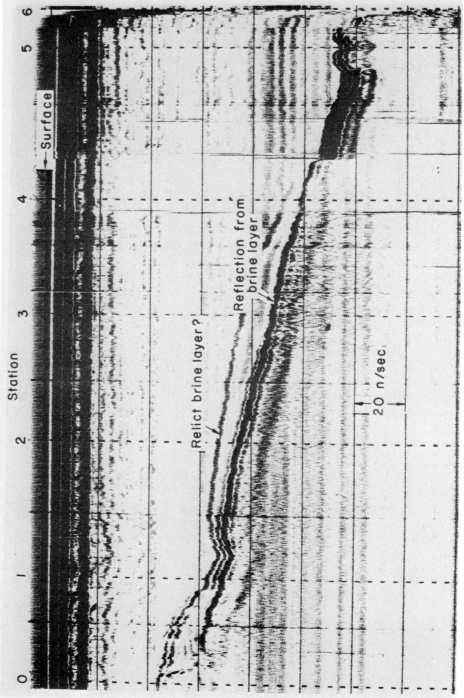

Fig. 85. Impulse radar profile showing multiple character of reflections from brine layer. The distance between stations is 305 m. (From Kovacs and Gow, 1975.)

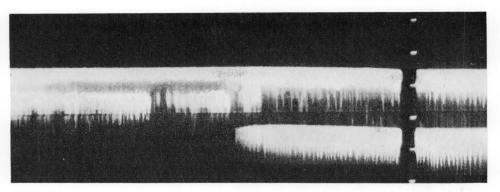

Fig. 86. Airborne radar profile across the edge of a presumed brine infiltration zone (left part of diagram) on the Larson ice shelf. (From Smith and Evans, 1972) Calibration marks are at 1 μs intervals. Reflections from the bottom of the ice shelf in the right part of the profile indicate an ice shelf thickness of 150 m. On the left of the photograph a reflecting layer has appeared at about 45 m below the ice surface and the bottom echo has been extinguished. Since the surface is flat Smith and Evans suggest that brine has percolated horizontally through the porous upper layers from one of the nearby rifts, which contain seawater.

Rayleigh fourth-power scattering law, i.e., a change in scattered loss of 40 dB per decade in radio frequency. The analysis showed that for bubbles of 1 mm radius or less, the energy loss due to scattering is not of practical significance at radio frequencies less than 1 GHz, but that the scattering might return a detectable signal.

A much more extensive look at scattering was undertaken by Smith and Evans (1972) [164], who considered mixtures of water, ice, and snow, using for the electrical properties of ice those values appropriate to polar glaciers. Considering first a surface meltwater layer, they show that the absorption loss in the water itself is negligible – not very different, in fact, from that in ice near the melting point: about 5 dB per 100 m. The principle loss comes from interference between reflections at the upper and lower surfaces of the water layer (Figure 87). For a water layer that is thin compared to the wave length, the transmission loss in dB for waves traveling through the layer $\simeq 2.4\, f_p h$ – approximately proportional to both the frequency, f_p, and the thickness, h, because the loss increases to reach a maximum as the thickness approaches a quarter wave-length. The maximum loss of 11 dB means that meltwater pools several meters deep on the surface will not cause a strong bottom echo to be lost. This is in accord with observations of airborne radar records by Swithinbank (1968) [167] and Smith (1972) [163].

Although the conductivity of rain water is relatively low, rain soaking could be important in some seasons because of the large thickness of a permeable temperate glacier that could be saturated. For soaked firn having a dry density of 0.5 mg m^{-3}, Smith and Evans (1972) [164] calculate a loss of 0.5 dB m^{-1}. If the dry density is 0.8 mg m^{-3}, then he loss is only about a third as great. This means that the bulk conductivity of rain-soaked snow is unlikely to produce serious difficulty in the sounding of temperate glaciers. Smith and Evans (1972) [164] point out, however, that the effect on velocity might be serious. Since the velocity in the soaked firn would only be about 40% of that in ice, 50 m of soaked firn could be interpreted as

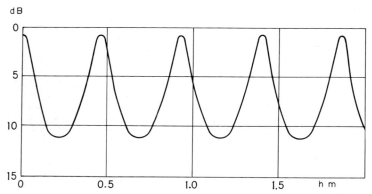

Fig. 87. The attenuation, in dB, of a 35 MHz wave passing through a layer of melt water ($\sigma = 10^{-4}\,\Omega^{-1}\,m^{-1}$) lying between air and solid ice, as a function of layer thickness in meters. In practice, such a layer is traversed twice and the total attenuation of the echo is twice the ordinate value. (From Smith and Evans, 1972.)

115 m of ice. This fact might be used to determine variations in water content from changes in travel time for reflections from the glacier bed.

Smith and Evans (1972) [164] next consider scattering losses, going beyond the case of air bubbles to consider holes filled with water. Fourth-power Rayleigh scattering still applies, but the attenuation is increased by a factor of 10 to 5×10^{-3} λ_0^{-4} dB/100 m for a density of $0.5\,mg\,m^{-3}$ (λ_0 is the free-space wavelength). The attenuation varies as the cube of the ratio of bubble size to wavelength. This still does not provide an important limit in the frquency to be used. As a limiting 'worst case', Smith and Evans (1972) [164] consider a very dense system of ice lenses, improbable in nature: low density snow with spherical ice lenses having radii of 50 mm and a spacing of 250 mm. The attenuation then increases to $0.7\,\lambda_0^{-4}$ dB/100 m.

Finally, Smith and Evans (1972) [164] show that, even for the worst ice lens case given above, the return from a planar reflector with a reflection coefficient of −20 dB would still be some 20 dB above the back-scattered noise. Nevertheless, they show examples where the bottom echo is clearly lost in scatter (Figure 88).

Some interesting intraglacial reflections were noticed by Goodman (1973 [83]; 1975 [84]) on Athabasca Glacier in connection with experiments using his 620 MHz radar. Soundings were made at the same location, under the same experimental conditions, with a time lapse of nearly one month between late August and late September. The results of the two data sets are shown in Figure 89. Although the bottom reflection remains unchanged, the structure of the internal reflections has changed markedly. Clearly these englacial reflectors cannot be associated with cracks, ice lenses, or changes in ice density. It seems probable that they are due to water layers forming and vanishing within the ice.

Strangway et al. (1974) [166] have studied interference patterns on Athabasca Glacier at several frequencies between 1 and 32 MHz, and compared them with theoretical curves (Section 4.2). At 8 MHz the field curves contain many more fluctuations than the theoretical curves. This is an indication of the influence of

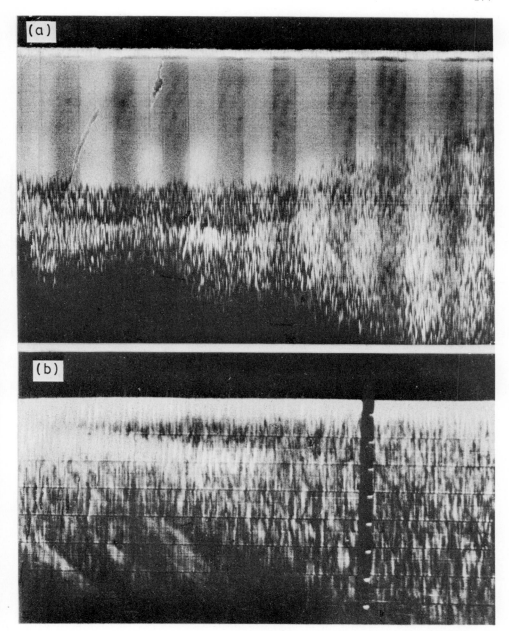

Fig. 88. Radar records showing the bottom echo disappearing into scatter echoes. The upper record (a), courtesy of Randall Electronics Ltd., was obtained on Hardangerjøkulen, Norway, using a 480 MHz sounder on the surface. The bottom echo, to the left, is at a depth of 100 m and it disappears into the scatter echoes to the right. The lower record (b) was obtained on the Fuchs Ice Piedmont, Adelaide Island, Antarctica using a 35 MHz sounder in an aircraft. The bottom echo is 40 dB above the receiver input noise level but it is, nevertheless, lost in the center of the picture at 150 m depth. (From Smith and Evans, 1972.)

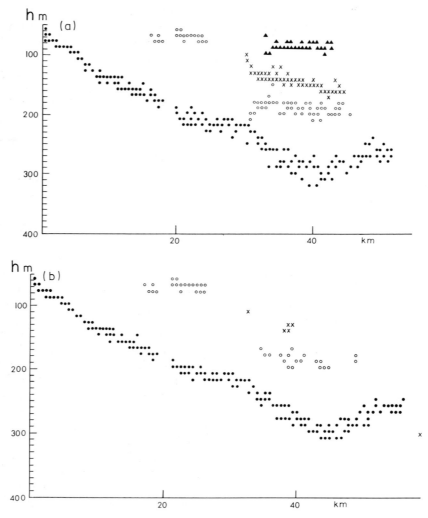

Fig. 89. Computer printouts of radar soundings made on Athabasca Glacier at two different times one month apart. The bottom echo is unchanged, but a marked change in the internal reflections can be seen. (a) Data tape taken 22 August. Weather sunny, no recent rain; (b) Data tape taken 21 September, about 60 cm snow on ice. (From Goodman, 1973.)

either inhomogeneities within the glacier or surface roughness. Scatterers that are considerably smaller than a wavelength will influence wave propagation very little, while those exceeding a wavelength in size will modify the fields considerably. The appearance of wavelength-size disturbances at 8 MHz (wavelength 38 m) indicates that internal ihomogeneities of the Athabasca Glacier are seldom larger than several tens of meters. A strong disturbance in all Athabasca data at 16 and 32 MHz (wavelength 19 and 9 m) indicates that scatterers of this size or less are common throughout the glacier.

One particular type of scatterer was investigated by means of a high-frequency analog scale model, using a styrofoam wedge to simulate a crevasse. The model field-strength profile showed an interference on the side of the crevasse nearer the transmitter and a minor diminution of field strength on the far side; similar features were seen in the field data. Because the dielectric properties of the model differ from those of ice, the results can only be taken as qualitative. Nevertheless, it seems likely that a crevasse caused the observed pattern, although the particular crevasse responsible was not identified.

EXPERIMENTAL STUDIES OF GLACIER STRATIFICATION

There are extensive internal layers in glaciers in addition to local inhomogeneities. These layers appear due to moraine and ice density changes. B.A. Fedorov, in the 11th SAE (1966), carried out introscopic glacier study along a 300-km track southward from Mirny Observatory by means of a RV-10 radio altimeter (Bogorodsky, Fedorov, 1967) [11]. These studies revealed three distinct echo groups. The first group includes pulses shaped by the interference between a sounding pulse and echoes from inhomogeneities close the ice surface. The second group includes echoes from deep inhomogeneities in the ice, and the third group includes bedrock echoes. When the data obtained along the track from Mirny were processed, the time delays were measured. Figure 90 displays internal inhomogeneities that caused the reflection of a sounding pulse. As can be seen in Figure 90, the maximum delay time decreases nearer to the coast. This seems to be attributable to a thinner layer with marked inhomogeneities (i.e., a thinner snow/firn layer) and an increase in glacier temperature. The latter (temperature increase) results in higher electromagnetic absorption losses, so that the reception of echoes from stratifications becomes worse. To make the detection of ice inhomogeneities better and to study the physical characteristics of the layers, radar sounding should be made at several frequencies (Bogorodsky, Fedorov, 1967) [11]. The frequency-dependence of the reflection coefficients for the observed layers makes it possible to subdivide the echoes into groups and to get better information about the physical properties

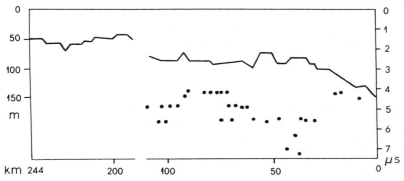

Fig. 90. Echo times for echoes from inhomogeneities in density and structure along a 300-km track from Mirny Observatory. The irregular line shows the snow–firn layer boundary; dots show the internal inhomogeneities.

of the reflecting objects than when equipment is operated at a single frequency. G.V. Trepov, in the 12th SAE (1967), employed soundings at two frequencies to study englacial structure – he operated his instrument at 100 and 440 MHz. The two radar receivers had identical sensitivity, with a pulse power of 130 dB. The study was performed along a track from Mirny towards Pionerskaya to a point 30 km from Mirny. Figure 91 shows the stratification close to the glacier bottom. Table XIX lists quantitative characteristics of the experiment (reflector depths, echo power).

It can be concluded from the comparison of sounding signals shown in Table XIX, and from echoes received from a single point at different frequencies, that the radar reflection pattern at a number of points (points 1, 5, 8, 10, 12, 16) is the same at both frequencies. This suggests that we are dealing there with the bedrock echoes. At other points (2, 3, 4, 6, 7, 9, 11, 13, 14, 15, 17, 21) echoes from layers close to the ice bottom were recorded. It should be noted here that the radar reflection patterns supplement each other. In the second case the last-arriving echoes were interpreted as bedrock echoes, whereas the earlier echoes can be regarded as reflections from moraine-containing ice layers located close to the bottom. The mismatch of bedrock echoes at the two frequencies at some points

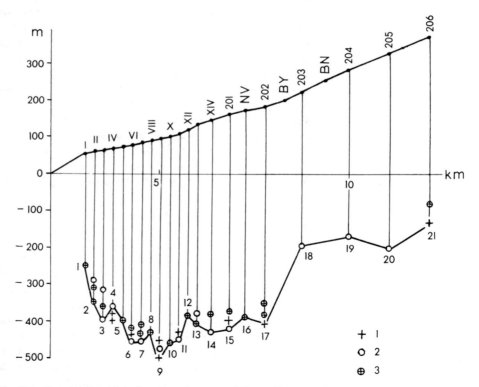

Fig. 91. Cross-section of the ice sheet along a track from Mirny to a site at the 12th km. Points denote the positions of reflections received (1) only when using the 100-MHz radar; (2) only when using the 440-MHz radar; (3) using both radar systems.

(points 3, 4, 13, 14, 17, 21) seems to indicate that the moraine-containing ice layer at those places is opaque to one of the frequencies, so that the internal echo could be misinterpreted as a bedrock echo.

The measurements at two frequencies were carried out only to the 12th kilometer. Further along the track ice the thickness was two great to be measured by a 440-MHz radar, so measurements were continued only at 100 MHz.

Figure 91, plotted from the data in Table XIX and other 100-MHz measurements, shows the surface and bedrock topography and the location of morainal layers. Morainal layers were found along most of the track.

There is some evidence to conclude that the distribution of moraine-containing layers is a result of ice movement over rock with this sort of topography. The height of these layers above the bedrock as measured by radar sounding (20–80 m) is in good agreement with visual observation of these layers in glacier profiles or icebergs that have been turned upside down. Experimental measurements at two or several resonant frequencies showed that such measurements greatly facilitate the interpretation of radar reflection data. Radar study of the glacier from Mirny to the 30-th kilometer showed the location of moraine-containing layers within the glacier and threw light on the bedrock topography.

Beginning in 1979, the possibility of using radar sounding to measure the thickness of annual snow layers has been investigated in the Soviet Antarctic Expeditions. The main difficulty in these measurements arises from the small thickness of the annual layer, ranging from a few centimeters in inland and marginal areas of the continent to 2 m 50–100 km from the coast. A radar system having a resolution of 1 ns is needed for this purpose, considering that the velocity of radio waves of centimeter and meter wavelengths in snow of density 0.4–$0.5\,\mathrm{mg\,m^{-3}}$ (this density is characteristic of the upper snow layers in the areas 10–300 km from the coast) is 210–$200\,\mathrm{m\,\mu s^{-1}}$. The necessary resolution can be obtained in the gigahertz range with pulse modulation. Attenuation of radio waves of this range in snow (10 GHz) never exceeds $1\,\mathrm{dB\,m^{-1}}$, so a relatively low-power radar, one with a gain of 110–120 dB, can receive echoes from depths greater than 20 m.

The electrical properties of interfaces within the snow are diverse due to different conditions of formation and metamorphism. Reflection coefficients at interfaces between snow layers of different densities are not large; e.g., for the boundary between snow with density $0.4\,\mathrm{mg\,m^{-3}}$ and snow with density $0.5\,\mathrm{mg\,m^{-3}}$, it is 0.029. Radiation crusts forming at the snow surface in spring and summer months have much higher coefficients. These crusts are less than a few millimeters thick, and they are normally underlain by loose snow of low density. The reflection coefficient for such a crust is 0.2 on the average, if the density of the upper part of the snow is $0.4\,\mathrm{mg\,m^{-3}}$ and that of the lower part is $0.2\,\mathrm{mg\,m^{-3}}$.

To measure snow thickness in Antarctica the radar described in Section 4.5 was used.

The equipment was mounted on a Kharkovchanka-2 vehicle; the antennas were 0.4 m apart and 1 m above the snow surface. Output signals, after gating, were fed into an intensity-modulated oscilloscope (model C1-65) in such a way that echoes from the surface and internal snow boundaries were displayed as a line of bright points that was photographed on a continuously moving film in the camera across the sweep line of the oscilloscope. When the Kharkovchanka-2 containing the mounted equipment moved, a radar profile of the snow layer was recorded on film (see Figure 92).

Interpretation of the radar profiles involved distinguishing and dating the characteristic boundaries. Figure 92 distinctly shows five reflecting boundaries, corresponding to the 1980, 1979, 1978, 1977, and 1976 layers. The four-year-mean snow accumulation in this area was estimated to be 145 cm yr^{-1}. This is in fairly good agreement with an average snow accumulation of 140 cm measured at stakes in the same area in 1970–1973.

To determine the radio-wave velocity in snow, and to check the interpretation of radar profiles, direct measurements of the depth of radiation and wind crusts were made in snow pits. Radar equipment was placed near the pits to make remote sensing measurements. To mark an echo from a measured boundary, a flat metal reflector was placed in the snow – this resulted in an increase of the amplitude of a measured pulse.

Measurements carried out at the 37th, 54th, 75th, and 105th kilometer of the Mirny–Pionerskaya traverse yielded a mean velocity of 207.5 m μs^{-1} with an rms deviation of 10.2 m μs^{-1}. Thus, the error due to velocity inaccuracy will be 5% of the measured snow depth.

This equipment also can be used for the study of the surface topography under snow in temperate latitudes, and for the detection of objects under the surface of the snow.

Radar sounding of polar ice sheets has revealed stratification to great depths, extending over hundreds of kilometers. The first observation of this in Greenland was made by Robin and others (1969) [153] during surface soundings in 1964. Intermittent echoes from a single layer were observed along the track from Tuto to Camp Century in north-west Greenland at depths ranging from 355 m to 610 m, and on a track south of Camp Century at a depth of about 440 m. They also reported on extensive layering at depths down to 2000 m observed in Antarctica during airborne soundings in 1967, later described by Harrison (1973) [100]. Other airborne soundings in 1969 (Gudmandsen, 1970) [88] showed a multitude of extended, continuous

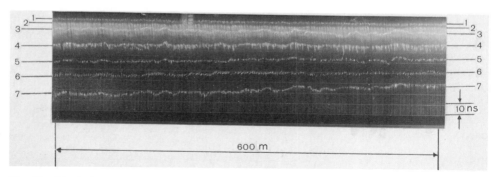

Fig. 92. Vertical radar cross-section of the snow–firn layer near the 37th km of the Mirny–Pionerskaya route. 1 – outgoing pulse, 2 – snow surface echo, February 1981; 3–7 – echoes from the tops of the 1980, 1979, 1978, 1977, and 1976 layers, respectively. $\rho = 0.4 \times 10^3$ kg m^{-3}, $c = 210$ m μs^{-1}.

TABLE XIX
Depth of the layer and echo power

No.	Stake	100 MHz		440 MHz	
		layer depth (m)	echo power (dB)	layer depth (m)	echo power (dB)
1	I	310	66	310	24;6
2	II	360	20		
		375	20	377	6
		420	26	420	6
3	III	394	26	394	26
		435	18	435	12
		455	18		
4	IV	435	44;6	455	6
5	V	465	12	485	2
6	VI	500	23	500	14
		535	6	520	3
7	VII	500	20		
		520	6	520	3
		545	6	545	6
8	VIII	530	19	530	3
9	IX	570	27	563;585	6;3
10	X	560	40	560	15;6
11	XI	545	6	555	2
12	XII	510	20	510	22
13	XIII	520	8		
		550	8	545	1
14	XIV	545	40	545	3
		585	35		
15	201	545	23	545	6
		585	20	570	3
16	HB	555	45;6	560	23
17	202	555	26		
		570	20	570	2
				595	2
18	203	420			
19	204	460			
20	205	545			
21	206	460	45	460	9

layers in central Greenland, at places almost down to the base of the ice sheet. The chances of observing layers are dependent, among other things, upon the performance of the sounding system, and the literature shows this clearly as more powerful equipment has become available – the most powerful being the RLS-60-74 radar referred to in section 4.3.

Since the first observation of internal layers, consideration has been given to the origin of the echoes and the mechanism that caused the layering. The one thing which may be said with certainty about the origin of layer echoes is that they are caused by changes in the permittivity of the ice. Such changes will cause a reflection, as we have already considered when discussing the capabilities of the measurement system and the possibility of detecting isolated layers of a certain thickness with a permittivity deviating from that of the surrounding ice. In this section we shall review briefly the various explanations put forward in the literature as to the cause of the reflections, i.e., variations in ice density, change in tan δ, changes in anisotropy, and ash and dust. Also, we shall consider the likely cause of the formation of the reflecting layers based on a combination of radar sounding and measurements on ice core samples. Robin and others (1969) [153] showed that small changes in ice density in an isolated layer yield reflection losses small enough to produce detectable echoes even with the low-performance equipment used in 1964 in northwest Greenland. Estimated reflection losses range from 43 dB to 80 dB in the area, where the ice temperature is about $-24°C$.

Since the refractive index may be expressed by

$$n \approx \sqrt{\varepsilon'} = 1 + (8.5 \times 10^{-4}\rho), \tag{6.13}$$

it follows that

$$\frac{\Delta\varepsilon}{\varepsilon} \simeq 1.7 \times 10^{-3}\Delta\rho, \tag{6.14}$$

where $\Delta\rho$ is in $kg\,m^{-3}$. Earlier in this section we derived a figure giving reflection loss versus layer thickness with $\Delta\varepsilon/\varepsilon$ as a parameter; (6.14) shows that this figure may also be used to estimate the reflection loss due to density variations. Thus, for a layer of relative thickness of 0.05 (0.14 m at 60 MHz, for instance), a reflection loss of 70 dB is obtained for $\Delta\varepsilon/\varepsilon = 0.002$ or $\Delta\rho = 1.2\,kg\,m^{-3}$.

It has been suggested that the change in density is associated with layers of clear ice surrounded by bubbly ice. Paren and Robin (1975) [138] show, on the basis of radar sounding data in central Antarctica, that this explanation may be valid to depths of the order of 1000 m. At greater depths the air bubbles are compressed by the weight of the overburden of ice, so that the difference between clear ice and glacier ice disappears. This was also stated by Harrison (1973) [100], based on data from ice coring at Maudheim, reported by Schytt (1958) [161].

A more or less extended and continuous layer of clear ice is likely to have been formed during periods of warmer climate. An attempt at correlating the layer echoes with the ^{18}O variations determined from Camp Century core samples was inconclusive, however (Gudmandsen, 1975) [89]. This makes the suggested clear-ice mechanism less likely.

Harrison (1973) [100] suggested that a variation in ice-crystal orientation in the layer relative to that in the surrounding ice could give the necessary change in permittivity. This suggestion was inspired by the observations by Jiracek (1967) [110] that showed the existance of dielectric anisotropy in the ice sheet of Antarctica. However, measurements at DYE-3 in Greenland at 300 MHz show a weak anisotropy that gives a gradual change in polarization from layer echo to layer echo between depths of 400 m and 1200 m. The change is proportional to depth (9.1 ± 0.3° per 100 m). This observation rules out the possibility of anisotropic layering. Harrison (1973) [100] also considered the importance of dust and ash layers as a cause for layer echoes. Such layers were recorded between 1300 m and 1700 m in the core drilled at Byrd station, Antarctica (Gow and others, 1968) [87]. The layers were about 0.5 mm thick. Considering the effect on the permittivity of mixed ice and rock particles, a reflection loss of about 110 dB was calculated, i.e., much larger than those observed with the radar sounder used – although still detectable by powerful equipment. Thus, that possibility also must be excluded.

On the other hand, the core sample showed that the dust and ash bands were always overlain by a layer of refrozen melt water which could attain a thickness of 10 mm. Reflections from these layers may just be detectable at the depth in question with powerful equipment. Paren and Robin (1975) [138] suggest that at greater depths, i.e., below about 1000 m, the layer echoes are due to changes in $\tan \delta$ relative to that in the surrounding ice. Previously, one of us (P.G.) also considered this possibility and found, in accordance with Paren and Robin (1975) [138], that this may give reflection losses of the same order of magnitude as those observed. This was substantiated by Millar (1981) [127], who analyzed recordings from 100 km northwest of Dome-C in Antarctica. Figure 93 shows the reflection loss for each individual layer as observed on the intensity-modulated recording.

Due to small-scale roughness, the return echo from a layer will fluctuate rapidly, within about 10 dB, as the sounder moves over the surface, so that averaging is necessary. This is done from A-scope recordings taken on film at intervals corresponding to 2 m along the traverse. As previously pointed out, determination of the reflection loss from a recording requires knowledge of the dielectric absorption in

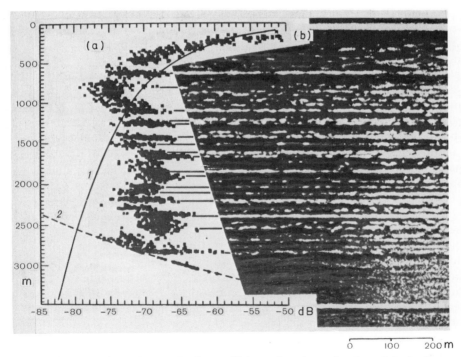

Fig. 93. On the left are shown reflection coefficients plotted as a function of depth: 1 – curve of expected losses due to layer reflections; 2 – the limit of layer detection using the 60-MHz radar. On the right the glacial layers are shown on an intensity-modulated display.

the ice overlying the reflector, normally calculated on the basis of an estimated temperature profile. One of us (P.G.) estimates that the individual measurements (indicated in Figure 93 by crosses) are correct within a 3-dB error that is due to uncertainty in the estimated temperature profile. Figure 93 also shows a curve that gives the expected reflection loss for a model of density fluctuations in layers of constant thickness (actually the thinning of the layers with depth should be taken into account; that would cause a shift of the curve towards larger reflection losses at greater depths). Finally, Figure 93 gives a curve of the detection limit with the 60 MHz equipment employed and the estimated temperature profile. Figure 93 shows two distinct regions: an upper one – to about 500 m depth – in which there is a rapid decrease in reflection loss with depth, and a deeper region in which reflection losses are within a 20 dB range centered around 70 dB. In the first region, reflections may be ascribed to density variations, whereas in the deeper region they are likely due to variations in $\tan \delta$.

Millar (1981) [127] refers to a suggestion by Hammer (1977 [91] and 1980 [92]) that the layer echoes are due to acidic layers, observed from ice coring, which may have a $\tan \delta$ variation sufficient to give such echoes. Hammer (1980) [92] finds on the basis of Gudmandsen and Overgaard (1978) [90] that the four most acidic layers found in the Crête core (Greenland) are all associated, within a ±5 m uncertainty in depth, with strong radio echoes. The acidity is determined by a current-probe technique applied to clean ice core surfaces. Dielectric measurements at 3 GHz on ice core samples, using a resonant method in which a dielectric cavity is formed by the core sample (Overgaard, 1981) [136], have shown a clear correlation between the two phenomena.

In view of the large horizontal extent of the layers, Robin and others (1969) [153] suggested that they were depositional in nature and that they may be used as time horizons (isochrons). Many authors dealing with this subject have addressed this point and derived the approximate age of a layer from information on surface velocities and surface mass balance. Now that the deep drilling to the glacier bed has been completed at DYE-3, Greenland (in 1981) an accurate dating of the layers can be made. The radar reflections will then permit an extension of the time scale to most of the Greenland ice sheet, to the benefit of ice dynamics studies. An example of such application of radar isochrons is given by Whillans (1976) [191] for Byrd Station, West Antarctica. Isochrons were calculated for 140 km along a flow line on the basis of measured surface velocity and mass balance and were found to agree fairly well with selected reflectors (2500, 5500, and 30 000 years old). (For a further discussion see Section 6.9).

However, to be able to use these isochrons safely, we should consider the accuracy with which they are defined vertically as well as horizontally. The first point has already been discussed in connection with the analysis of the measuring system. It was found that the individual layer depth may be determined with an accuracy of about 15 m. With the present state of ice dynamics modeling this is a sufficient accuracy. However, if one wants to correlate radar sounding data with those obtained from an ice core, the task is far more difficult, the depth accuracies being at least two orders of magnitude different. This statement is related to Z-scope representation, which is normally used for isochron studies. The situation may be improved, if A-scope data are used instead to resolve the pulse length/bandwidth dependence.

Harrison (1973) [100] raised the question of whether the layer echo observed was due to an isolated single layer or was caused by a sequence of layers within the pulse resolution length (one-half the pulse length in the ice medium). In comparing data obtained with a 250 ns pulse to those obtained with a 1 μs pulse, he found that the echo strength with the longer pulse was 5 dB larger, which could be understood as an integration effect, since the strength of echoes from multiple layers can be shown theoretically to be proportional to the pulse length. However, Clough (1977) [55] points out rightly that the reflection from a single rough surface will show the same pulse-width dependence, since the reflecting area involved will be proportional to the pulse length.

Figure 94 resolves this question to the satisfaction of both authors. It shows a Z-scope recording from Antarctica with application of different pulse lengths in a rather homogeneous region. With the 1 μs pulse, the resolution length is about 85 m, and it is seen that vertical integration clearly takes place so that the number of layers observable is reduced compared with the cases of 250 ns and 60 ns pulses. No integration seems to take place with the two shorter pulses for which the resolution lengths are about 21 m and 5 m respectively, since the number of layers is unchanged from one pulse length to the other.

Also, Figure 94 shows that the vertical integration which takes place with a 1 μs pulse creates an uncertainty in the position of a particular echo, and that the continuity of a layer is broken from time to time. However, this is probably due to the display techniques employed, in which the intensity-modulated signal is differentiated before being displayed on the Z-scope. The intensity of a reflection is there-

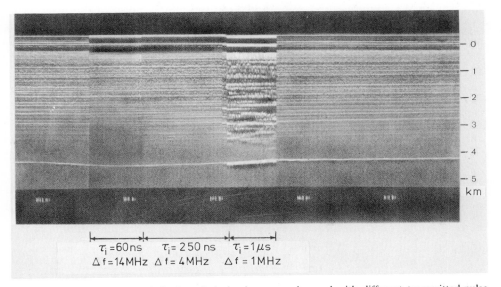

Fig. 94. Intensity-modulated display of glacier layers as observed with different transmitted-pulse widths, τ_i, and receiver band widths, Δf. Changes in resolution can be seen clearly.

fore a function of the rise time of the pulse, and the trailing edge of the pulse will create a negative signal that may cancel or reduce sharply a subsequent echo. With a reflection loss for an individual layer that may vary by more than 10 dB, as shown in Figure 93, a shift from one set of layers to the subsequent one, or a change in the integration process between the layers, may take place.

Finally, Figure 94 shows the pulsewidth/bandwidth effect already referred to – it appears as a shift in the layer's position by about 12 m from the 60-ns pulse to the 250-ns pulse recording, and a corresponding shift by 60 m from the 250-ns pulse to the 1-μs pulse recording, when disregarding the layer definition alluded to above. Figure 94 also shows the improvement in sensitivity obtained by reducing the receiver's bandwidth. With the 1 μs pulse (1 MHz bandwidth) a layer is observed at a depth of 3765 m or 440 m above the glacier base, whereas the deepest layer with the 60 ns pulse (14 MHz) is at 3295 m (and, in fact, is observable only because of visual extrapolation from that recorded with the 250 ns pulse (4 MHz).

These considerations are important to the horizontal continuity of the isochrons, which may suffer from the variability in reflection loss as described. Fortunately, the gross structure of the layering does not change rapidly from place to place, so that re-establishment of 'lost' layer echoes often may be accomplished visually. It is clear, however, that when automated layer tracing is attempted by computer, operator intervention may be needed from time to time in an interactive analysis procedure.

In mountainous regions, the form of the surface reflects the bottom relief, and it is found that this is also the case for the layering. The layers 'follow' the form of the bottom, but the layer variations are increasingly damped or smoothed as the surface is approached. In principle, this observation may be used to throw light on the flow of the ice over the bottom relief but, unfortunately, disturbances are introduced from the sounding procedure and from the modification of the wave propagation by the undulating surface. Harrison (1971) [97] studied the geometry of echoes from a stack of continuous, undulating layers observed by a sounder with a wide downward-looking beam (as in the 60 MHz system previously described). He was able to describe a number of sounder observations of different layer configurations, and found that the appearance of a set of these layers depends on the height of the sounder relative to the center of curvature of the undulating layers. The layer echoes will create cusps and points of strong signal at positions that depend on the layer configuration – in fact, the position of a cusp defines the curvature of the layer surface at that place. Due to interference between echoes that differ in range by less than a pulse length, a fading will be observed; the fading pattern is different for the cusps and the cross-over points of strong signal. In addition, since the ice surface also will be an undulating surface, it may act as a lens (due to refraction through the surface) and give a focusing and defocusing effect for the convex and the concave parts of the surface, respectively. This also will create a fading of the layer echoes as the sounder moves over the surface.

Since the geometry is known, it is possible in principle to reconstruct the actual layer configuration from the sounding records, taking the surface refraction into account. An example is given by Harrison (1971) [97] who shows a rough reconstruction based on a slope correction and the knowledge of the curvature of the reflecting surface when a cusp is recorded. The procedure requires detailed sampling of the recorded signal and automated (computer) handling of the data. Without such a procedure great care must be exercized in drawing conclusions from recordings of layers in mountainous regions.

A disturbing effect is encountered in Greenland in areas where summer melting frequently takes place. The melt water penetrates the snow surface and refreezes into ice lenses. These lenses form efficient scatterers in the firn and, due to the broad antenna beam, they reflect appreciable energy at comparatively long distances from the airborne sounder. In certain areas the effect is so strong that the relatively weak layer echoes cannot be detected – in some cases even the reflection from the bottom may not be discernible due to this surface 'clutter'. A contributing effect is the large dielectric absorption produced by the relatively high ice temperature that predominates in the same areas. In fact, summer melting takes place occasionally over large parts of the Greenland ice sheet, so that surface echoes are discernable in most recordings.

6.9. Movement of Glaciers

The study of glacier movement is an important aspect of glaciology. The problems of glacial dynamics must be solved if one is to understand the interaction mechanisms in the ice – ocean–atmosphere system. Ice velocity can be now measured by means of radar sounding and laser techniques, which give higher acuracy than conventional geodetic and astronomical methods or repeated mapping of glacier thickness.

Noteworthy also is a technique of indirect ice velocity estimation which implies a comparison of surface slopes with bedrock topography. It is based on a theoretical assumption that surface slopes are functions of ice movement velocity.

HORIZONTAL MOVEMENT FROM THE RELATIONSHIP BETWEEN ICE THICKNESS AND SURFACE SLOPE

It has long been known that profiles of the ice and rock surfaces show maximum ice-surface slopes occurring directly above peaks in the rock surface (Bourgoin 1956 [41]; Robin, 1958 [147]; Govorukha et al., 1974 [14]). Although this correlation agrees qualitatively with the basic Nye theory that relates surface slope to basal shear stress (Nye, 1957 [130]), the quantitative agreement is poor.

Variations in surface slope are commonly much larger than those predicted by Nye's theory. This was explained by Robin (1967) [148] as being due to variations in the previously neglected longitudinal stress. Backed by a more rigorous theory by Collins (1968) [57], Robin showed that variations in surface slope around a regional mean can be related to longitudinal stress gradients which, in turn, are related through the flow law to the longitudinal strain rate. Since the horizontal velocity, in turn, is simply the integral of the strain rate (where there is no basal sliding), a quantitative comparison between surface slope variations and detailed bedrock topography determined by radar sounding yields a measure of the horizontal velocity. Robin (1967) [148] used this theory to show that a better fit was obtained to the surface topographic variations along a line south of Camp Century, Greenland, by using a velocity that was half that expected for a steady-state ice sheet, rather than by using a steady-state velocity (Figure 95).

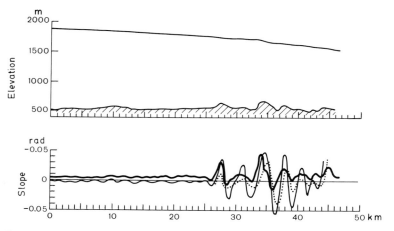

Fig. 95. Comparison of observed (heavy continuous line) and calculated surface slopes along an estimated flow line extending 46 km southward from Camp Century, Greenland. Light continuous line: slopes calculated for estimated 'balance velocity'. Dotted line: slopes calculated for a velocity equal to half the balance velocity. The upper diagram in the figure shows the ice thickness profile along the same line. (From Robin, 1967.)

Beitzel (1971) [32] carried out a similar analysis in Antarctica, although he assumed a velocity appropriate to a steady-state ice sheet and took as the adjustable variable the exponent in an assumed power flow law for ice. Presumably, as the flow law becomes better determined, this procedure can become increasingly effective in determining movement rates where more direct measurements are not available.

Budd (1971) [42] has improved the theoretical treatment by applying spectral analysis to bedrock and surface profiles, finding that Robin's (1967) [148] formula needs modification by a factor of 2, thus probably bringing the surface slope variations into agreement with a steady-state ice sheet. Budd and Carter (1971) [43] show futher that the theory explains the relationship between bedrock and surface profiles in Queen Maud Land (measurements by Beitzel, 1971) [32] and Wilkes Land. The technique of determining the flow parameters for longitudinal strain-rate from spectral analyses of bedrock and surface profiles along a flow line, where the flow is sufficiently two-dimensional and the velocity is known, appears to be quite accurate and sensitive. Armed with such knowledge of the flow parameters, it will become possible to use the same technique to calculate unknown velocities in other areas. Radar sounding is necessary for this approach to provide the bedrock topography in the detail needed for spectral analysis.

MEASUREMENTS OF ICE MOVEMENT VELOCITY BY MEANS OF A LASER INTERFEROMETER

Laser interferometry is a very promising technique. It is based on the Doppler effect. It is possible to determine the velocity of a reflector rigidly fixed to the ice surface by changes in the beat frequency of a coherent signal. A description of velocity measurements on Antarctic glaciers by laser and radar equipment (Belousova et al., 1971 [3], Bogorodsky et al., 1976 [9]; Bogorodsky, Fedorov, 1970 [12]) follows. (The laser interferometer was designed by I.P. Ivanov; Dr. I.M. Belousova supervised the work.)

Measurements of ice velocity by means of the laser interferometer described in Section 4.7 were carried out in the 15th SAE (1970) and in the 18th SAE (1973). Experimental polygons were located at the margin of Hays outlet glacier in the vicinity of Molodezhnaya Station and near Vetrov Hill in the Mirny Observatory area, where ice movement is extremely slow. Ice strain rates and their relationships to different ice morphology were studied at other polygons situated 50 km from the coast. The glaciers chosen for experimental measurements were fairly well studied – their mean annual velocities were known and their subglacial topography had been mapped by radar. Ice velocities at the edge of Hays Glacier were measured near Vetrov Hill in January and February, 1973. Measurements were made at

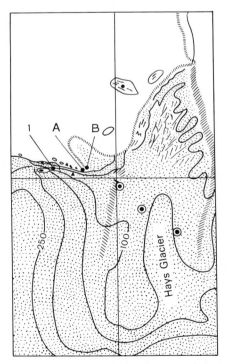

Fig. 96. Map of the coastal-end polygon for ice velocity measurements at the periphery of Hays Glacier, showing 1 – Mount Vechernaya; dots A and B – laser-interferometer measurement points; circled dots – sites of velocity measurement by stake displacement.

the end of the melt season when the ice surface was bare of snow. Antarctic outlet glaciers are characterized by a large amount of dynamic crevasses criss-crossing their surfaces.

Hays Glacier moves over its bedrock with a 5° slope. It terminates with a 20-meter-high barrier in the Alasheyev Bight, Kosmonaut Sea. The bedrock, as found by seismic and radar soundings, slopes gently seaward. At sites A and B (Figure 96) ice velocities were measured to vary, within a minute, from 0 to $20\,\mathrm{m\,s^{-1}}$. Such irregular ice movement led to the introduction of the term 'instantaneous velocity'.

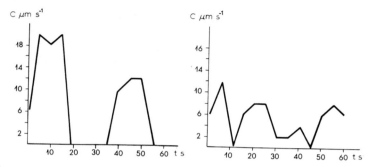

Fig. 97. 'Instantaneous velocity' measurements on Hays Glacier for two different one-minute observation periods.

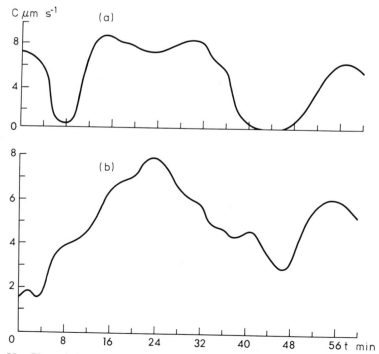

Fig. 98. Plots of the 'minute-velocity' over an hour at sites A and B on Hays Glacier.

The 'instantaneous velocity' is defined as the velocity averaged over one second, a definition justified by experimental measurements that showed that, during one second, the ice velocity changes very little. A few hundred measurements of the instantaneous velocity were made at each of the two sites. About 80% of all the measurements showed the instantaneous velocity to change arbitrarily during a minute (Figure 97). In other cases no ice movement was detected. No regularity in ice velocity changes was found in these experimental surveys, hence a chaotic character of 'instantaneous velocity' changes in the marginal area of Hays Glacier can be considered typical. The 'instantaneous velocities' were used to determine the velocity averaged over a minute; changes in this 'minute velocity' in the course of an hour were then plotted (Figure 98). The 'minute velocity' of Hays Glacier was found to range from 0 to 8 μm s^{-1}, though velocity variations occurring during an hour were smoothed. Average hourly velocities were found to vary randomly within the observation period. Table XX lists the most frequently occuring (i.e., modal) ice velocities of Hays Glacier at sites A and B.

TABLE XX
Average hourly velocities

Date (1970)	Area A					Area B				
	30/1	31/1	31/1	1/2	2/2	29/1	29/1	2/2	3/2	4/2
Velocity (μm s^{-1})	3.6	5.6	4.7	4.6	4.8	5.8	5.0	5.2	3.2	2.5

It follows from Table XX that average hourly velocities range from 3.6 to 5.6 μm s^{-1} in area A and from 2.5 to 5.2 μm s^{-1} in area B. The ice velocity for this region thus can be estimated to be approximately 120 m yr^{-1}. Velocities determined earlier in this region by other methods varied from 120 to 480 m yr^{-1}. It should be emphasized that Hays Glacier was found to have a complicated pattern of movement, stopping or drastically changing its velocity. Its 'instantaneous velocity' varied from 0 to as much as 20 μm s^{-1}. The mean hourly velocities, however, fluctuated only slightly. Thus, for large time intervals, Hays Glacier seems to have a translational velocity of 0.3 m day^{-1}.

Glaciers with small velocities cover 98.5% of the Antarctic ice sheet. To measure the velocity of such slow-moving glaciers, a glacier near Mirny Observatory was chosen. The ice velocity and its changes with time at various points in a polygon were measured, using Vetrov Hill as a reference. The polygon was 1500 m long and 80 m wide, with a rough surface that had an average slope of 10°. The elevation change over a distance of 1.5 km was 40 m. The ice velocity measurements had been carried out several times in the Mirny area – they showed the velocity to be 0.4–0.7 m yr^{-1}. The first experimental laser surveys indicated a velocity two orders of magnitude lower than that for Hays Glacier mentioned above, with a fluctuation period of about 10 s. Thus, for the calculations of averaged 'minute velocities', mean velocities for 10 s were used. More than a thousand measurements were made. They demonstrated that the ice velocity in the area did

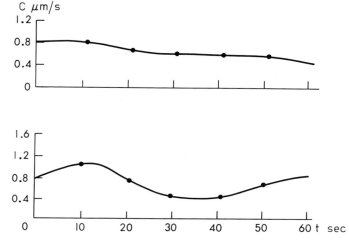

Fig. 99. Variations in 10-second-average ice velocity over two one-minute periods near Mirny.

not change significantly, all values of velocity being almost identical. The average hourly velocities were determined by the 10-seconds measurements. Figure 99 shows typical fluctuations in the 10 second average velocity over two one-minute periods, near Vetrov Hill. Estimated hourly average velocities are listed in Table XXI – they range from 0 to 1.2 μm s^{-1}. The ice velocity in the area was found to depend mainly on the temperature, so velocity measurements were carried out at the same time, that is from 1600 to 1800 local time. For each of the points under study, fluctuations of the 'minute velocity' during an hour were first determined, then average hourly velocities were calculated. On the basis of these values, daily variations of the ice movement velocity were plotted (Figure 100). Then the relationship between glacier flow and distance from the outcrops was studied (Figure 101).

TABLE XXI

Average hourly velocities of points at Mitny Polygon (in the vicinity of Mirny Observatory)

Date	Point number	Velocity (μm s^{-1})	Date	Point number	Velocity μm s^{-1}
23/2/1973	1	0	5/3/1973	10	0.3
27/2	2	0.1	18/3	11	0
1/3	3	0.4	3/3	12	0.1
3/3	4	0.6	14/3	13	0.3
4/3	5	0	6/3	14	0.7
5/3	6	0.1	14/3	15	0.8
5/3	7	0.3	18/3	16	1.2
3/3	8	0	4/3	17	0.4
4/3	9	0.1			

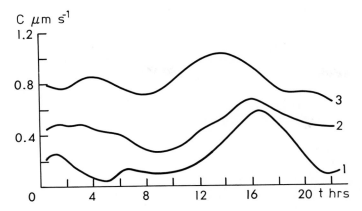

Fig. 100. Daily variations in hourly-average ice velocity near Mirny on 1–20 February, 1973; 2–28 February 1973; 3–18 March 1973.

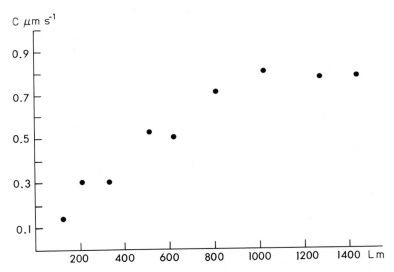

Fig. 101. Velocity of various points on a glacier near Mirny in relation to distance from outcropping rock.

Figure 101 indicates that there is an inactive zone near obstacles where the ice velocity is equal to zero, then for some distance from the obstacle there is a region of growing ice velocity, and finally there is a region of steady velocity. Figure 101 also shows that the increase in ice velocity with distance from the nunataks is from 0.1 to 0.7 $\mu m\,s^{-1}$. At a distance of 500 m the ice velocity increases almost linearly. The annual glacier movement estimated by the results of the laser measurements ranges from 0.3 to 2.4 m, whereas the velocity of the same glacier calculated by stake data ranges from 0.2 to 1 m yr^{-1}.

The ice velocity was measured along a transverse profile on the same glacier. The reflectors (measuring points) were emplaced 100 m apart along the profile. This profile, with eight measuring points, was 1000 m from Vetrov Hill. At that distance nunataks do not affect the ice flow, which was therefore assumed to be steady. The measured curve of average hourly velocities is almost sinusoidal. Thus, ice velocities are the same in magnitude, but different in direction due to the angle difference. The equality of velocities and symmetry of angles enabled the general direction of ice movement to be determined on a base line 1000 m long. It was found to coincide with that of a point with the highest velocity.

HORIZONTAL MOVEMENTS FROM FADING PATTERNS

The basic procedure in this method is to record the shape of the radar echo returned from the rock bed, i.e., to record the amplitude as a function of time until the end of the echo. If the instrument is moved horizontally through the diffraction pattern formed by interference between reflecting facets on the rough glacial bed, the shape of the returning pulse will change, and large differences will occur in horizontal distances of the order of the wavelength of the radiation, say 5 m. Since the diffraction pattern is determined by details of the reflecting rock bed, it should remain fixed in position while the ice moves. It thus forms a fixed reference against which ice movement can be measured in places far removed from visible landmarks. Analog experiments carried out in the laboratory by Nye *et al.* (1972) [133], using ultrasonic waves in air in place of radio waves in ice, suggested that it should be easy to detect a movement as small as a tenth of a wavelength and, perhaps, with further refinement, even less.

In particular localities it could be that the signal apparently returning from the bottom of the ice is partly or wholly due to scattering by a layer of rock debris entrained in the ice near the bottom. If this layer were moving relative to the true base of the ice, then the apparent movement at the surface would reflect the difference in movement between the surface and the deep reflecting layer rather than the true movement of the ice across its bed. Such a layer might easily be formed by a concentration of moraine in the basal layers of the glacier. In this case, the velocity measured would result from strain within the ice. Comparison with the total velocity determined either by surveying techniques or by accurate satellite position measurements would give the part of the velocity attributable to a combination of strain in the ice below the reflecting layer and sliding on the glacier bed – principally the latter, since the morainal material is likely to be concentrated very close to the bed. This is a very exciting possibility, since there is no other known method for differentiating between the sliding and integrated-strain components of ice movement.

One other possibility also needs to be considered. If the diffraction pattern were made up of substantial contributions both from the base of the ice and from one or more deep internal reflectors, then it would be substantially modified in the course of time as the glacier moved and might, therefore, very soon become unreproducible. To the extent that this were true, there would be a time limit within which the repeated measurements needed to be made before all recognizable characteristics of a particular diffraction pattern were lost.

Field experiments were carried out on the Fleming Glacier (Antarctic Peninsula) in January 1972 using a SPRI Mark IV sounder operating at 35 MHz (Walford, 1972) [183]. Fleming Glacier was picked because it is a deep, cold glacier, and was expected to be fast-moving. A particular location was chosen where the glacier surface was smooth and crevasse-free, and a strong fading pattern had been observed from airborne observation. The experiment was carried out during a single visit.

Eight lines, each 7.5 m long and forming a fan-shaped pattern, were surveyed. The fading pattern along each line was recorded daily for 3 days (Figure 102). Corresponding features and comparable fading patterns were measured and the component of glacier velocity along each line calculated and plotted (Figure 103). These components should lie in a circle whose diameter represents the true velocity of the glacier. A least-squares analysis gave $139 \pm 10 \, \text{m yr}^{-1}$ for the velocity of the glacier. By conventional optical surveying techniques, the velocity of the glacier is $168 \pm 2 \, \text{m yr}^{-1}$ (C.W.M. Swithinbank, personal communication quoted by Walford, (1972) [183]), with an azimuthal difference between the two measurements of about 15°. On the basis of the precision attainable in his three-day experiment, Walford (1972) [183] did not feel justified in concluding that there was a significant difference between the velocities determined by the two methods.

In 1974, Doake (1975) [63] visited a nearby site on the same glacier in an attempt to achieve greater precision in both the radar and optical surveys. Fading patterns were recorded over a series of parallel lines along a calibrated track, and a series of five measurements was made over a period of one week. The average velocity found was $138 \pm 5 \, \text{m yr}^{-1}$. An optical survey was carried out by intersection on six reference objects on surrounding rock outcrops. The velocity calculated from these

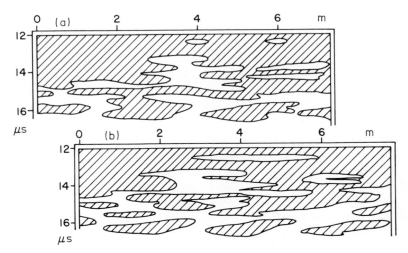

Fig. 102. Radar fading patterns obtained on Fleming Glacier, Antarctica at 1106 GMT on January 28, 1972 (upper), and at 1242 GMT on January 31, 1971 (lower) along a line bearing 267° from a marker fixed in the snow. The data were derived from records photographed at 25 cm intervals along the line. Hatched areas show where the echo strength is high. The site was displaced by approximately 1 m in 3 days. (From Walford, 1972.)

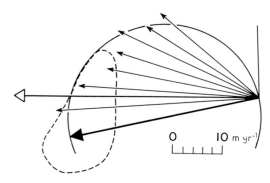

Fig. 103. The velocity of Fleming Glacier, Antarctic Peninsula. Light arrows represent measured velocity components in indicated direction. The heavy solid-headed vector arrow is the vector resultant. The open-headed vector arrow denotes the velocity determined by conventional optical surveying techniques. The circle shown best fits the radar results; the dashes outline the 95%-confidence region. (From Walford, 1972.)

observations by two different methods gave a speed of $200 \pm 20 \text{ m yr}^{-1}$. There is clearly a significant difference between these two velocities, which Doake interpreted as being due to basal sliding or rapid deformation of a basal reflecting layer. Doake confirmed his measurements on a return to the same site the next season (Doake, 1975) [64], obtaining a substantial improvement in the accuracy of the optical survey (movement rate $201 \pm 4 \text{ m yr}^{-1}$), and confirming the speed indicated by the radar technique also, although with somewhat lesser accuracy ($142 \pm 11 \text{ m yr}^{-1}$). Furthermore, he found that higher up the glacier the displacements determined by the radar method were the same, within the limits of error, as those determined by optical survey. Presumably in that region the reflecting surface was a true glacier bed and not an imbedded morainic layer.

Another application of the method was made on the Devon Island Ice Cap in Canada (Doake et al., 1976) [65]. Here the temperature at the ice–rock interface was known to be $-18.5°C$ from data in boreholes that reached the bedrock, so that no sliding could be expected. Because the boreholes were near the ice divide, the velocity would be expected to be very small. A total of 150 measurements were made at a point about 2.5 km downslope from the borehole. Although the basal temperature might be somewhat warmer at this point, it could hardly be above the pressure melting point. The radar survey point was connected to the boreholes at the ice divide by surface markers that permitted the surface strain rates to be measured and compared with the radar measurement of velocity.

The equipment used to this case was the SPRI Mark IV 60 MHz system with a $1 \mu s$ pulse length. Compared with the $0.3 \mu s$ pulse length used by Walford (1972) [183] and Doake (1975) [63]; (1975) [64], this system could be expected to give bottom echoes of longer duraction and, therefore, spatial patterns with a wider range of fading lengths, thus giving more detail to be used as a basis for comparison.

The measurement procedure was essentially the same as that of Doake (1975) [63]. The antenna was supported a few centimeters above the ice on a 'railway' constructed of glass fiber tubing. The antenna was then moved steadily along the railway and the fading pattern recorded. Repeated traverses were made with the antenna shifted laterally each time by an average of 0.2 m. In this way, the fading pattern was mapped over an area of 8 m by 2 m.

Fading patterns were obtained in June 1973 and June 1974. A total of 69 individual comparisons gave a movement of $2.58 \pm 0.03 \, \text{m yr}^{-1}$. The error figure refers only to the comparison of the fading patterns and does not include any possible systematic error. An estimate of these leads to an error figure of $0.11 \, \text{m yr}^{-1}$.

On the Devon Island ice cap it was not possible to measure movement directly by optical techniques. Instead, the strain rates were determined and integrated to determine the velocity at the radar measurement site, assuming that the velocity is 0 at the ice divide. That assumption was checked by inclinometer measurements at the borehole near the divide. The strain rate measurements yielded a velocity determination of $2.17 \pm 0.20 \, \text{m yr}^{-1}$.

Although the velocities measured by the two methods do not differ significantly at the 90% confidence level, the difference is large enough to suggest that a real physical explanation should be considered. In contrast with the situation on Fleming Glacier, on Devon Island the velocity determined by the radar technique was higher. The probable reason is that part of the strain line did not coincide with the flow line in the ice, so that the measured strain rate was less than the greatest principal strain rate, leading to too low a calculated velocity. In this case, the radar method is probably the more accurate.

Soviet radio physicists started their ice movement studies and ice velocity measurements from the development of radar equipment. Radar specifications are discussed in Section 4.6. The same section describes some physical aspects of radar measurements of ice velocity. The general physical basis of all the experiments carried out by the AARI specialists is similar to that described in detail at the beginning of this section. To study the characteristics of spatial fading, G.V. Trepov and B.A. Fedorov measured fluctuations of an echo at various ice thicknesses. The signal amplitude with parallel polarizations of receiving and transmitting dipoles would normally change with ice thickness and temperature; in addition it will change randomly due to unknown factors. Trepov started the study of echo fluctuations in the 16th SAE (1971). Recording was made at three frequencies (60, 213, and 440 MHz) from a slowly-moving vehicle over sections of track ranging from 20 to 100 m long in various experiments. The recording rate was 200–250 frames for each 100 m of track. Echo amplitudes were recorded, since the display was of the A type. To study the stability of the spatial fading and the dependence of its statistical characteristics (in particular, the auto-correlation radius for amplitude fluctuations) on ice thickness, measurements were made at several sites with different ice thicknesses. Since an electromagnetic signal changes its polarization while propagating through the ice, and is also controlled by the bedrock morphology, the radar was moved along two perpendicular routes, one north–south and the other east–west.

Figure 104 shows the dependence of echo amplitude on position. Computer processing of the amplitude measurements with a preliminary normalization factor

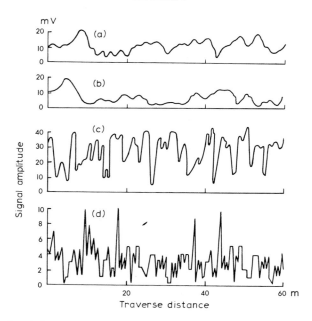

Fig. 104. Fluctuations in bottom-echo strength: (a) and (b) at 60 MHz for two mutually perpendicular directions yielding R_a = 5.2 m and 5.1 m, respectively; (c) at 213 MHz, giving R_a = 1.6 m; (d) at 440 MHz, giving R_a = 0.8 m. R_a is the autocorrelation radius.

made it possible to determine the auto-correlation radius, R_a. For 60 MHz it was found to be 5.1 ... 5.2 m, while for 213 and 440 MHz it was 1.6 and 0.8 m, respectively. The scattering properties of the bedrock at the indicated frequencies were found to be similar along various track directions. The lower the frequency and the larger the ice thickness, the more stable the fading pattern seemed to be.

The experiments and their analysis made it possible to improve the accuracy of the formula proposed by Watt and Maxwell (1960) [185] for the estimation of the spatial fading distance.

In the 21st SAE (1976) Trepov and Sheremetyev (Bogorodsky, Trepov, and Sheremetyev, 1979) [10], using a 60-74M radar, measured the velocity at 19 points along the line AB near Molodezhnaya Station (Figure 105). First, the pattern consisted of 5 lines diverging from a common center, each 16 meters long. The line and step lengths for the survey were chosen to ensure reliable recording of the fading-pattern. The mean values of maximum and minimum spatial fading distance are known to be determined by the following expressions:

$$L_{max} \approx \frac{\lambda\sqrt{2}}{4}\sqrt{\frac{h}{0.4\,\tau_t c_i}}; \quad L_{min} = \frac{\lambda\sqrt{2}}{4}.$$

Knowing that, for the 60-MHz radar, λ = 2.8 m in ice and τ_t = 0.5 μs, and that h_{max} = 4000 m, gives L_{max} = 11 m, L_{min} = 1 m. Thus, displacements of the ice surface relative to the bedrock should be measured over track sections that are shorter than L_{max} and longer than L_{min}.

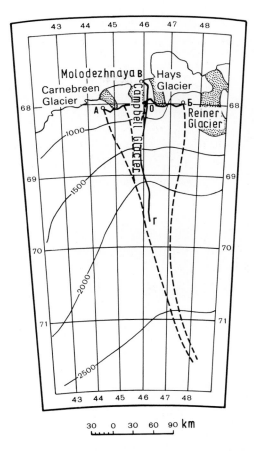

Fig. 105. Location map for velocity measurements in the area of Carnebreen, Campbell, and Hays Glaciers.

The central radiating arm of the polygonal recording pattern coincided with the direction of the steepest surface slope, and the angular distance between two adjacent arms was 10 to 15°. In each survey the line of tractor-sledge motion relative to the ice surface was marked by a tight wire. Film motion for photographic recording of the echoes was synchronized with the tractor's motion, a recording being made every 20 cm along the line of travel.

The displacement of the ice surface was determined by the displacement of the fading patterns between the beginning and the end of the survey. The direction of movement was assumed to coincide with the azimuth of the arm along which the displacement was the largest. Displacements measured at 19 points ranged from 0.2 to 1.6 m, corresponding to velocities of 8–208 m per year. Figure 106 shows examples of records of fading patterns (Carnebreen Glacier). A displacement of 1.2 m took 118 h, which corresponds to a glacier velocity of 89 m y^{-1}. Figure 107 shows velocities and speeds (moduli of the vector velocities) calculated by radar

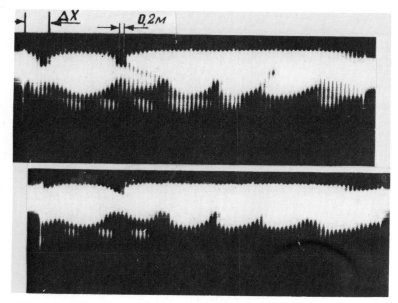

Fig. 106. Measurement of ice velocity on Carnebreen Glacier; ice thickness is 800 m.

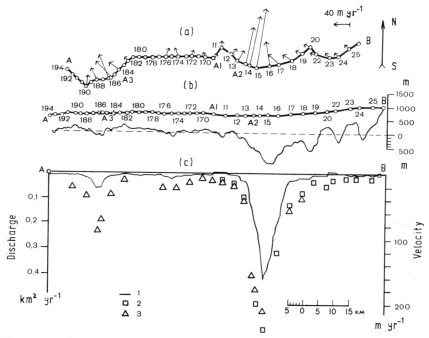

Fig. 107. Profile of ice velocity along line AB in Figure 105. a – velocity vectors; b – vertical cross section of the ice; c – solid discharge measured in longitudinal area of a column crossing a unit length of AB, per year; 1 – calculations of solid discharge; 2 – speeds using geodetic data obtained by GDR scientists in 1972, 3 – speeds determined from radar sounding results of 1976.

measurements along the AB track. Speeds measured by geodetic methods are also shown in this figure.

The direction of ice movement was found to lie within ±15° of the slope of the ice surface. The error value was determined by the accuracy of the method of measuring direction.

Ice thicknesses and ice velocities determined by the radar sounding technique in the Molodezhnaya Station area were used to estimate ice discharge from the basins of three outlet glaciers: Hays, Campbell, and Carnebreen Glaciers. The results are shown in Figure 107.

In the 23rd SAE, the same techniques were used to measure ice movement along the track from Mirny Observatory to Pionerskaya Station, using a 60-74M radar mounted on a Kharkovchanka-2. Faster recording equipment made it possible to shorten the time needed for each polygon to 15–20 minutes. Figure 108 shows the results.

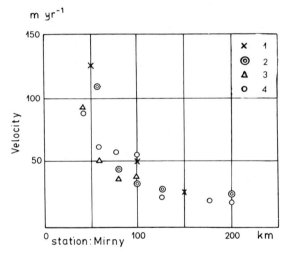

Fig. 108. Ice velocity along the track Mirny–Pionerskaya. 1 – estimated; 2 – from astronomic observations; 3 – from geodetic data; 4 – from radar sounding results.

Geodetic and astronomic measurements of ice velocity in this area preceded those of radar sounding. The geodetic measurements were carried out by scientists from the German Democratic Republic; the astronomical measurements were made by an Australian scientist using a Geoceiver, which analyzes Doppler variations in signals transmitted by US navigational satellites. Ice velocities were also estimated using an expression relating ice thickness, ice surface slope and velocity given by Budd (1975: Russian Translation [2]):

$$v = \frac{2h^{n+1}}{n+1}\left(\frac{\rho g \alpha}{B}\right)^n, \tag{6.15}$$

where v is the ice velocity in $m\,s^{-1}$; h is the ice thickness in m, α is the ice surface slope angle, ρ is the ice density in $mg\,m^{-3}$, g is the acceleration due to gravity

(9.81 m s^{-2}), and B and n are the parameters of the ice flow law, for which the values $B = 0.68 \times 10^9$ Pa s$^{-1/3}$ and $n = 3$ were chosen.

The results are shown in Figure 108, which shows a fairly good agreement between measurements made by different methods.

Ice thickness and velocity should be known in order to study flow laws for large ice masses moving from outflow centers to the coast. These quantities were measured in the austral summer of 1979/80 during the 24th SAE along the track from Mirny to Komsomolskaya to Dome B. A map of the area is shown in Figure 109. Measurement results are listed in Table XXII. The velocity was found to vary from 120 m yr^{-1} in coastal areas to 1 m yr^{-1} in inland Antarctica.

To estimate the accuracy of the radar technique, the ice velocity was measured near a stake whose displacement was measured by a Geoceiver. These measurements (54 km from Mirny) indicated the ice velocity to be 117 ± 9 and 121 ±

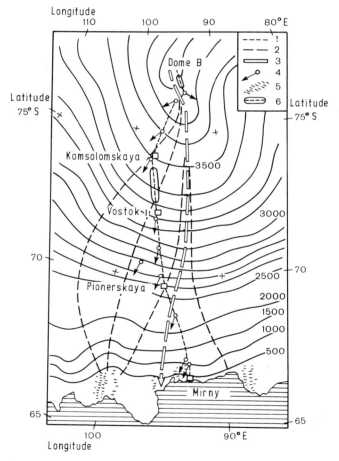

Fig. 109. Map of the region where measurements were made along the track Mirny–Dome B. 1 – track of the scientific traverse; 2 – ice divide; 3 – presumed flow line; 4 – velocity vectors; 5 – outlet glaciers; 6 – areas where basal melting was found.

TABLE XXII
Velocity of ice movement in East Antarctica

Distance from Mirny (km)	Position of the polygon	Slope of the surface (radians)	Glacier displacement (cm)	Time (h)	Glacier velocity (m yr^{-1})	Azimuth of the movement (deg)
54	67°00'S 93°17'E	0.0146	130 ± 10 310 ± 20	97 224	117 ± 9 121 ± 7	20
73	67°10'S 93°20'E	0.0122	200 ± 20	220	78 ± 8	20
252	68°40'S 94°31'E	0.0117	160 ± 20	793	18 ± 2	20
390	69°52'S 95°46'E	0.0026	100 ± 20	676	13 ± 3	33
480	70°40'S 96°01'E	0.0020	40 ± 10	627	6 ± 1.5	0
615	71°52'S 96°24'E	0.0013	30 ± 10	547	4.8 ± 1.6	357
872	74°07'S 97°30'E	0.0011	260 ± 20	8424	2.7 ± 0.2	30
965	74°56'S 96°38'E	0.0022	120 ± 20	8352	1.26 ± 0.2	40
1070	75°49'S 95°22'E	0.0023	100 ± 10	8354	1.05 ± 0.1	65
1150	76°26'S 93°55'E	0.0003	80 ± 10	8328	0.74 ± 0.1	300

7 m yr^{-1} by means of radar sounding, and 123 ± 8 m yr^{-1} as calculated from position changes (Young, 1979) [196]. This close agreement suggests that englacial morainal material hardly affects the echoes in this area.

Figure 110 presents examples of fading patterns obtained during velocity measurements near Dome B.

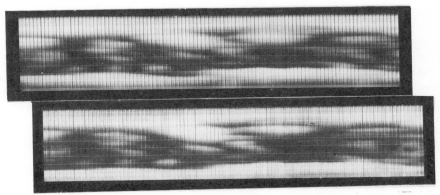

Fig. 110. Velocity measurement in the vicinity of Dome B. Ice thickness is 2100 m. 1 m of ice displacement in 348 days was obtained on a track 20 m – this corresponds to a velocity of 1.05 m yr^{-1}.

VERTICAL MOVEMENT FROM FADING PATTERNS

In principle, the fading pattern produced by diffraction effects in the echo from base of a glacier varies vertically as well as horizontally. However, on a thick ice sheet, the pattern varies rapidly horizontally but is relatively insensitive to vertical displacement.

To make the fading pattern sensitive in the vertical direction, Nye et al. (1972) [132] proposed mixing the returning echo with a continuous wave at the carrier frequency. The shape of the mixed signal will bear critically upon the phase relationship between the two carrier waves. If the distance between the antenna and the bed of the glacier changes by a quarter wavelength, an in-phase point on the return signal will be changed to an out-of-phase point. Thus, the shape of the mixed signal is sensitive to displacement in the vertical direction.

In the technique proposed by Nye et al. (1972) [132] an observation point on an ice sheet would be carefully mapped at some time and then reoccupied at a later time, say one year or more later. Since the purpose of the experiment is to measure vertical changes, a location would be chosen where the horizontal movement would be relatively small. Upon return to the observation site, a point lying in the same vertical line as the original observation point could be found by observing the horizontal fading pattern. Thereafter, one could, in principle, relocate the original observation point by moving the antenna vertically until the shape of the original echo was recovered. In practice, it would be better to keep the position of the antenna fixed and to change the carrier frequency enough to recover the original shape of the mixed echo. The relative change in height above the bed, H, is numerically equal to the relative change in frequency, so that a 1 cm change on an ice sheet 3000 m thick would correspond to a change in frequency of 120 Hz for a 35 MHz carrier. It follows that no extraordinary frequency calibration is demanded. Because the pattern repeats itself when H changes by one-half wave length, that is about 2.5 m, the period between observations should not be so long that there is any ambiguity in identifying the true vertical displacement. Alternatively, any ambiguity could be resolved by repeating the observations using more than one frequency.

There are several difficulties associated with this scheme. It is, of course, important to measure any phase shifts that might arise from a change in the performance of the electronic equipment between two observation times. The principal instrumental limitation, however, is the signal-to-noise ratio. Nye et al. (1972) [132] point out that an accuracy of 10 mm in the height change using a wavelength of 5 m would require a signal-to-noise power ratio of 32 dB, so a strong bottom echo is required. The signal-to-noise ratio could be improved by digital signal averaging techniques.

Another difficulty that Nye has considered in detail (Nye, 1975) [131] is the effect of irregularities in the accumulation rate. Since the velocity pattern in an ice sheet will change very slowly with time, responding only to a very long-term average change in the accumulation rate rather than the short-term changes that are actually observed, short-term changes in the height of the surface will occur. Nye suggests minimizing this problem by taking as a reference point a marker at a certain depth below the surface rather than the surface itself.

An additional problem that is mentioned by Nye et al. (1972) [132], but not considered further, is that the procedure they recommend neglects any effect due to a change in the average wave velocity throughout the ice sheet over the period of observation. This could be a significant factor. Since the number of cycles of the carrier wave in one round trip through the ice is inversely proportional to the wave speed, the measurement is just as sensitive to changes in wave speed as it is to changes in height. Thus it would only require a relative change in mean velocity of 3×10^{-6} to give the appearance of a 10 mm height change on an ice sheet with a constant thickness of 3000 m. The corresponding mean density change would be 6×10^{-6} mg m^{-3}. Relating this to the uppermost part of the ice sheet, the corresponding change would be, for example, 0.06 mg m^{-3} spread over only the top 0.3 m of the ice sheet. Such a change is well within the normal range of variations in the uppermost layers of the firn. Thus accurate density–depth information would need to be collected at both times of measurement. (The change in density in the upper part of the ice sheet that would be required to correspond to an apparent change in height is independent of the ice thickness.)

The initial survey for one experiment to measure vertical movement has been made on the Devon Island ice cap (Walford et al., 1977) [184]. Radio echoes were measured repeatedly, and precisely, at single sites near the crest of the ice cap. Each site was surveyed by theodolite with respect to a local gravity station that is to be maintained for a number of years. Radar spatial fading patterns were recorded locally and measurable markers were buried in the snow. The amplitude and phase of the reflections were recorded as functions of time over networks of sites. The relative height of the antenna at each point within the network was known. A separate survey site was also set up for future re-measurement to determine horizontal displacements.

The experiments are described in great detail by Walford et al. (1977) [184]. Careful auxiliary meausrements were made, including the physical and electrical length of each section of coaxial cable used, and the speed of the direct signal propagating just above the surface between transmitting and receiving antennas. They also found that the phase change with height above the surface was somewhat less than would be expected for a specular reflector, a fact that they attribute to the large cone of illumination (35°) and the roughness of the bed.

INTERNAL ECHOES AND THE MEASUREMENT OF PALEOVELOCITIES

A new, and potentially powerful, approach for studying the present and past dynamics of large ice masses involves the use of internal reflectors. The method appears first to have been applied by Robin, Evans, and Bailey (1969) [153]. There are two principal assumptions involved: that an internal reflector was formed at a time when the reflecting horizon was at the surface of the ice sheet, and that the vertical strain rate along any vertical line in the ice is uniform at any instant. It then follows that the vertical distance between internal reflectors will diminish with time in proportion to the diminishing height of the layers above the bedrock. Robin et al. (1969) [153] then proceed to calculate the depth at which a particular layer would be found, depending on whether the ice sheet is in steady state so that its thickness

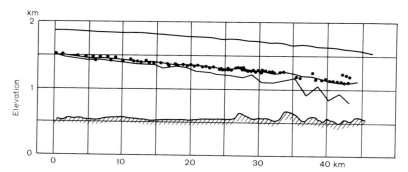

Fig. 111. Reflecting horizons along the profile south from Camp Century, Greenland. The dots mark individual reflector positions measured from the radar film. The lines are based on the two methods described in the text. (From Robin, Evans, and Bailey, 1969.)

is a constant at any point or, as another extreme, is not in motion at all, so that the surface elevation is increasing at the surface accumulation rate. The results of these two calculations compared with observations are shown in Figure 111. It is immediately apparent from the figure that the steady state model provides much the better fit along the whole profile. The calculated depth according to this model differs from the observed depth by less 6% at all points, and the agreement at the end of the profile (between 40–43 km) is within 2%. The procedure then can be taken as a confirmation that the velocities appropriate to steady state, i.e., 3 m yr^{-1} at Camp Century increasing to 7 m yr^{-1} at the 30 km point and 23 m yr^{-1} at 44 km, are not seriously in error. (The fact that this conclusion contradicts that of Robin (1967) [148], cited earlier in this section, is not commented upon by the authors.)

Whillans (1976) [191] has applied a modification of this method to a section of the West Antarctic ice sheet upstream from Byrd Station, where surface strain rates have been measured. His method involves the use of surface mass balance, surface strain rates, and the form of the internal layers as determined by radar sounding, but does not involve the use of ice thickness and bottom topography directly, although, of course, the strain rate and velocity as measured at the surface are, in part, physically determined by them. The procedure is to compare the variations in depth of the internal layers, as measured, with those calculated on the basis of models of ice flow. A very simple model for the flow of the ice sheet was used by Whillans (1976) [191]. He assumed that strain rates, as measured at the surface, are constant throughout the thickness of the ice mass, but that they vary horizontally. Thus, the horizontal velocity of a vertical column of ice is the same at all depths. The depth at any later time of a layer initially deposited on the surface may thus be calculated.

Whillans (1976) [191] measurements and calculations showed agreement that is much too good to be merely coincidental (Figure 112). This shows that the simple ice flow model is a good approximation for the flow of the ice sheet, supports the concept that the reflecting layers are connected with the depositional stratigraphy and therefore represent isochrons, confirms the conclusion that, at present, this part of the ice sheet is almost in a steady state, and, finally, indicates that this

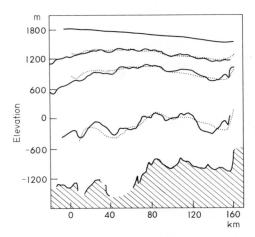

Fig. 112. Profile along the axis of the strain net from Byrd Station to the ice divide. The radar reflectors are shown by the solid lines and the calculated isochrons by the dotted lines. Approximate ages of the isochrons are 2500, 5500, and 30 000 years, respectively. Stippled area represents bedrock. (From Whillans, 1976.)

portion of the ice sheet has been close to its present form for a long time, probably more than 30 000 years. Because of the extensive occurrence of internal echoes in large ice bodies, the method should have widespread applicability as the number of regions where the requisite surface measurements have been made increases. As the method is refined, it should be possible not only to test for the presence or absence of the steady state, but also to produce a quantitative estimate of past variations in ice movement.

Robin *et al.* (1977) [152] have used internal reflectors to test the assumption, made by Whillans (1976) [191] and by Robin *et al.* (1969) [153], that there is uniform vertical strain in a vertical column. Choosing three layers from a radar profile in central East Antarctica with estimated ages of 52 000, 113 000, and 180 000 years, they show that, despite inaccuracies in following individual layers, in 90% of the measurements the height of the ice column between the internal layers varies in the same proportion (to within ±10%) as that between the surface and the uppermost layer. Between the lowest reflector and bedrock, however, the thickness of the layer varies by a factor approaching twice the mean value. In this lowest layer near the bedrock the shear stresses will be dominant, whereas in the higher layers longitudinal stresses will cause most of the deformation. In the upper layers, which are colder and much more rigid, then, the ice moves as a slab which deforms uniformly throughout to match changes in ice thickness. In contrast, the layer above the bedrock varies considerably in thickness along the flow line (Figure 113). It is hoped that this type of information from radar sounding can be used to develop useful modifications of the theories of ice flow.

Not only is the lowermost layer variable in thickness, but it appears to be devoid of internal reflectors (Figure 113). Considerations of the system performance and of the possible nature of flow around a rough bedrock topography led to the sugges-

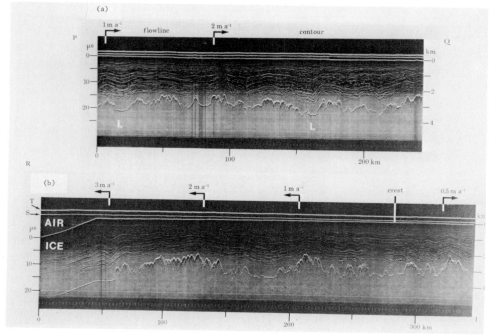

Fig. 113. (a) Radar profile showing bedrock and internal layering near Dome B in central Antarctica. The section on the left lies along a flowline, that on the right shows the profile along a surface contour of the ice sheet. (b) Radar profile across the crest of the ice sheet. The profile lies along a flowline and shows changes in deformation of layers as the ice movement increases away from the crest of Dome B. (From Robin et al., 1977.)

tion that the layering is not visible because it is no longer present due to the variable deformation over a rough relief. This concept calls in question the validity of dating ice cores close to bedrock, since uniform deposition and strain are used in calculating dates at these levels. Where continuity of internal layering is clear, there is little doubt that age continuity down the core is satisfactory. However, where there is no layering present and yet the system performance is adequate in relation to absorption so that layering would be seen if it were present, serious doubts about continuity arise. Not only are such observations useful as a check against the validity of age determinations, but as system performances improve enough so that very weak echoes in the basal non-reflecting zone can perhaps be found, more information about the nature of the ice deformation and flow in the presence of high subglacial relief should be obtainable.

Another variation on this procedure is presented by Whillans (1979) [193]. The model used in this case is based on surface velocity and net surface accumulation rates, and also on the measurements of tilting in the deep drill hole at Byrd Station, which are used to estimate the rate of change of vertical strain rate with depth. This model differs from the ones previously described in that the assumed constant vertical strain rate is replaced by one that varies by about a factor of 2 from the surface to the bottom, in accordance with actual measurements. At present the ice

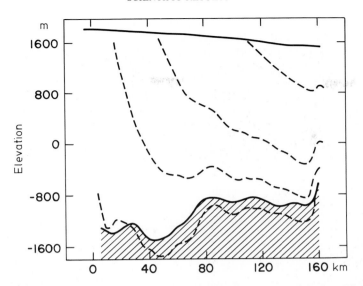

Fig. 114. Internal particle paths for the West Antarctic ice sheet near Byrd Station. The positions of particle paths (dashed lines) are plotted with respect to the present-day surface elevation of the ice sheet. The surface of the shading represents a running mean, over 12 km, of the position of the bed as determined by radar sounding. The deepest particle path is also the calculated bed position: its shape is similar to that obtained by radar but calculated thicknesses are greater, indicating that the ice sheet was thicker in the past. (From Whillans, 1979.)

sheet is thinning slowly near Byrd Station, as shown by the analysis of surface data and radar measurements of ice thickness (Whillans, 1977) [192]. This rate is slow enough not to contradict the conclusions drawn from analysis of the internal echoes (Whillans, 1976) [191], since it does not imply a rapid thinning. Nevertheless, the thinning rate is taken into account in the present analysis. Using these data, Whillans calculates particle paths along the Byrd Station flow line (Figure 114). The deepest particle path shown in the figure indicates a calculated position of the bed along the flow line. The good agreement between calculated and radar-determined bed shapes lends support to the velocity model assumed. Unfortunately, Whillans (1979) [193] does not show calculations based on different assumptions, so that it is not clear how closely the ice velocity field can be delimited by this technique.

GLACIAL FLOW LINES

In previous sections we have seen how ice shelves are characterized by several different kinds of reflection phenomena relating to the nature of the basal interface and to inclusions within the ice. It is possible to take advantage of these characteristics to study the movement of the ice by tracing them across the ice shelf. Such studies have been applied in some detail to the Ross ice shelf and have led to a considerable advance in our knowledge of the velocity vectors, both past and present.

Neal (1979) [129] was the first to do this, using the occultation of the bottom return by a brine infiltration layer (cf. Section 6.2), reflection coefficients at the ice–water interface (cf. Section 6.10), and one abrupt ice thickness change of 30–40 m to draw flow lines across the grid eastern part of the Ross ice shelf. Flow lines drawn on that basis are generally in close correspondence with velocity vectors that have been determined by direct measurements. In a like manner, Bentley *et al.* (1979) [37] correlated sections of strong, weak, and intermediate reflection strengths (Figure 115) and associated them with flow lines, also on the grid eastern part of the ice shelf (Figure 116). The main features of their flow line map agree closely with those on the map presented by Neal. It is clear that strong reflectors are associated with ice flowing out of the main outlet glaciers through the Transantarctic Mountains. Particularly prominent are zones associated with the Beardmore, Nimrod, Byrd, and Lennox-King Glaciers. In addition, there are some strongly reflecting zones associated with valley glaciers along the Transantarctic Mountains that do not drain ice from the interior plateau. One, in particular, can be traced as a bright band partly across the ice shelf, but then becomes weaker and narrower and disappears as one approaches the seaward margin of the ice shelf. Further along the flow line the band shows additional appearances and disappearances on a time scale of hundreds of years. Bentley *et al.* (1979) [37] suggest that the alternate appearances and disappearances correspond to past variations in the activity of the alpine glacier with which the band is associated.

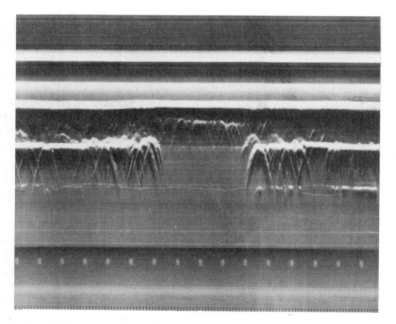

Fig. 115. Section of airborne radar profile on the grid eastern part of the Ross ice shelf, showing sections of strong bottom echoes (lower thick white line) separated by a zone in which the bottom echoes are obscured. Echoes from a presumed brine layer in the central part of the ice show weakly but clearly. (From Jezek, 1980.)

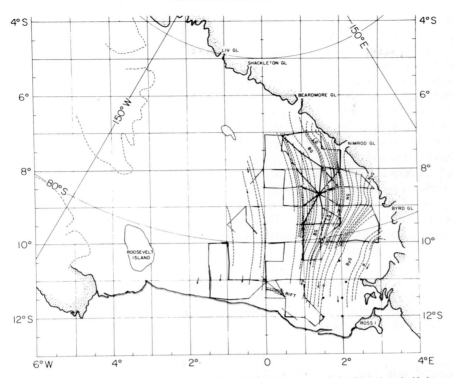

Fig. 116. Map of flow lines (dashed lines) on the grid eastern part of the Ross ice shelf drawn by correlating radar reflecting and non-reflecting bands. Solid lines show flight-line control; circles represent surface stations. (From Bentley et al., 1979.)

Variations in the mesoclimate could cause a reduction in the flow from the glacier to the point where cracking through the ice and brine infiltration could take place.

The study of the bands associated with valley glaciers in the Transantarctic Mountains was extended by Bentley (1981) [36], who examined in more detail a particular group of valley glaciers that form four distinct bands between Beardmore and Nimrod Glaciers, both of which are major East Antarctic outlet glaciers. That group was selected principally because of the large number of radar sounding tracks (about 25 flights; see Figure 116) crossing their associated flow bands.

Two characteristics of the glacier bands were measured, the width and the reflection strength. In order to make an approximate correction for the down-stream convergence of ice flow lines, the widths were measured as percentages of the total width of the band from the grid eastern edge of Beardmore Glacier ice to the grid eastern edge of Nimrod Glacier ice. No quantitative measure of the reflection strengths was available, so instead a qualitative assessment was made on a five-point scale (excellent, good, fair, poor, and non-existent, assigned numerical values of 4, 3, 2, 1, and 0, respectively). Upon careful examination of the echograms, it became clear that there were, in three of the four bands, two sets of echoes, either one or both of which might be present in a particular location: the principal echo from the bottom of the ice shelf, and an echo from an internal reflector at a height of roughly 50 m above the ice–water boundary. Reflection strengths were assigned separately to the two sets of echoes, since, to a considerable degree, they vary independently of each other. Finally, a semi-quantitative 'quality' factor was assigned to each reflector at each crossing of each glacier band by multiplying the relative width by the numerical reflection strength.

A scale of residence time of the ice shelf was developed using direct observations of ice movement. 'Quality' was then plotted as a function of residence time. Each of the valley glacier bands show pronounced downstream variations. The bottom echoes for all four glaciers have been relatively low in 'quality' for the last 600 years. Prior to that there are peaks of high 'quality' for each glacier lasting 50 to 150 years, but the correlation between bands is rather poor. For the internal reflector, however, the inter-band correlation is much better. All three bands show very similar patterns: high 'quality' since about 300 years ago (the record begins at about 200 BP), between 500 and 800 PB, and more than 1150 BP (the record ends at 1350 BP), with periods of very low 'quality' in between. These variations are not minor – 'quality' factors along a particular band vary by an order of magnitude or more.

Just what causes these striking variations is not at all clear. It seems inescapable that some sort of variation in the valley glaciers themselves has been occurring over the last 1500 years, but it is difficult to be more specific.

A particularly striking feature of the record is the virtual disappearance of both basal and internal reflections associated with the glaciers about 500 years ago. This might be attributable to rifting and brine infiltration above the firn–ice boundary, as is believed to occur between glacier bands, although there is no direct evidence of such an occurrence. However, most of the variations are not attributable to that mechanism, because the basal or near-basal echoes, even when weak, are clearly present.

An extensive study of the ice flow on the other side of the Ross ice shelf has been made by Jezek (1980) [107]. In contrast with the situation in the grid eastern part of the Ross ice shelf, mapping flow lines on the grid western part is not a straightforward task. The bottom reflection is nearly uniform and in only a limited number of cases can changes in the reflection strength or reflections from internal debris in the ice be correlated between flight lines. Fortunately, there are considerable variations in the strength of the multiple reflection (i.e., the reflection that has made two round trips through the ice), which fall within the dynamic range of the recording system and so are easily seen.

Jezek (1980) [107] mapped these variations over the entire ice shelf (Figure 117), and found good correspondence with the results of the other previously-mentioned study carried out over the grid eastern sector of the ice shelf. In agreement with Neal (1979) [129], he divides the ice shelf into three regions. Bright returns on the grid easternmost region fall into well-defined bands. The whole central section of the ice shelf is generally a poor reflector, excepting the intermittent bright bands that flank the outflow around Crary ice rise. This region encompasses all the outflow from Ice Stream A and much from Ice Stream B. Ice multiples in the grid western part of the ice shelf show more complicated patterns. The returns are

SCIENTIFIC RESULTS 215

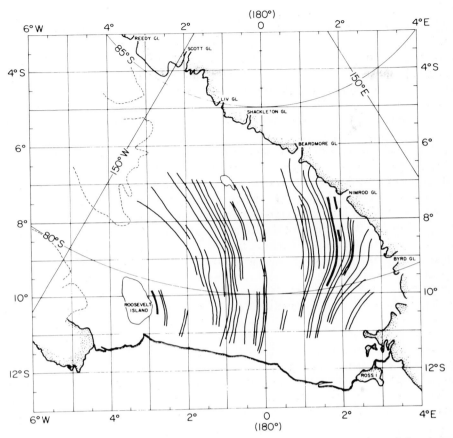

Fig. 117. Flow lines on the Ross ice shelf based on the correlation of multiple-echo variations between flight lines. (From Jezek, 1980.)

generally brighter and the width of constant strength echoes along individual flight lines is also greater.

In a few cases the 'debris track' of a scatterer embedded in the ice was traceable downstream for long distances. Primary sources of such tracks, presumably, are brine layers and included moraine. Particularly interesting examples appear downstream of Crary ice rise. The most easily identified is a strongly-absorbing layer which begins immediately downstream of Crary ice rise and appears clearly on many of the flight lines. Within this band is a strong reflector that, unlike most other radar targets in the area, starts near the bottom of the ice shelf and then rises and brightens as it moves downstream. Jezek (1980) [107] believes that this is caused by the inclusion of moraine into the ice shelf at the grounding line of Crary ice rise.

A noteworthy feature of some of the debris tracks is a significant divergence from the present-day flow directions in the ice shelf. This is strikingly so for the bands downstream of Crary ice rise and for another band downstream from the margin of ice stream D. If, as seems likely, the debris tracks are caused by the injection of material into a point at or near a grounding line, then their divergence from the present-day flow lines may be an indication of past movement of the grounding line. Jezek (1980) [107] has used this approach to conclude that the size of Crary ice rise diminished rather rapidly between 700 years and 400 years ago, and that the ice shelf must have been thinning at that time. Such an interpretation is also consistent with the debris track stemming from the margin of ice stream D, although in that case the situation is less clear because of the uncertain position of the grounding line. An alternative explanation for the divergence, that the flow vectors have changed in the past, Jezek considers less likely because of the absence of similar divergences in the grid eastern section of the ice shelf.

Jezek (1980) [107] has carried this analysis further to include an explanation for some of the striking ice thickness variations indicated on the radar thickness map. In particular, he identifies maxima and minima downstream from Crary ice rise as having formed in the vicinity of the ice rise, and then having broken away from the ice rise between 400 and 700 years ago as the ice shelf thinned.

Jezek's (1980) [107] interpretations are open to question and further study, but they show dramatically the utility of radar studies for examining the glacial history of the Ross ice shelf in the last 1000 years or so. Presumably the same sort of study could be made on other ice shelves.

6.10. Nature of the Basal Interface

Information concerning the nature of the basal interface from radar sounding alone is derived primarily from the strength of the bottom echo. There are two principal characteristics of the basal interface that affect the bottom echo: the impedance contrast and the roughness of the interface. The theoretical aspects of the influence of each of these factors on the reflection coefficient are discussed in Section 3.7.

The most fundamental characteristic to be determined from reflection amplitudes is whether the ice rests on water or rock. A signal reflected from a grounded bottom may be reduced by 10–30 dB, whereas the loss at a saltwater interface is less than 1 dB. In principle, this large difference should make it easy to tell from airborne radar surveying where the boundary between grounded ice and a floating ice shelf occurs. In practice, however, many such boundaries in Antarctica have not been readily identified. There are several explanations for this failure, including non-specular reflection, the possible existence of a layer of saline ice at the base of ice shelves, and the fact that in many places there appears to be a zone of substantial width between full-grounded ice and fully-floating ice, within which grounding occurs in zones or patches on a wet bed. Nevertheless, in some instances reflection amplitudes have been useful in distinguishing between grounded ice and ice shelves and in identifying water bodies beneath the inland ice sheet.

In principle, the amplitude of the reflection from the base of grounded ice should give information about the nature of the subglacial material, even though variations from rock type to rock type are not as great as variations between rock and water. In practice, analyses of this type have not been carried out extensively, probably because of the difficulty in accounting for other factors that affect the reflection amplitude. Absolute reflection coefficients are much more difficult to measure than relative amplitude changes from place to place because it is necessary to calculate correctly the losses due to spreading and to absorption within the ice. The former are easy to calculate, although minor uncertainties arise where the velocity-depth gradient is not well known, but uncertainties about both the temperature in the ice and the absorption coefficient as a function of temperature can lead to substantial potential errors in estimating absorption losses. In the future, as it becomes possible to calculate such other effects more accurately, the reflection coefficient should be used more frequently to determine the nature of the subglacial rock. Table XXIII (from Jiracek, 1967 [110]) gives a list of selected basal materials with their associated reflection coefficients and corresponding reflection losses at the ice/material boundary. Dielectric constant, conductivity, and loss tangent values are also shown. Many of these values can only represent rough estimates at 30–300 MHz, since the measurements were made at other frequencies. Also, the majority of measurements were taken at temperatures well above freezing.

TABLE XXIII
Reflection losses at a boundary between ice and other materials

Material	σ_∞ ($\Omega^{-1} m^{-1}$)	ε_∞	Reflection loss (dB)
Freshwater	5×10^{-3}	88	3.4
Saltwater	3	88	0.5
Sea Ice	1.1×10^{-3}	3.4	36.0
Rocky Land	2×10^{-3}	10	11.0
Gravel	3×10^{-3}	3.2	22.3
Quartz	10^{-4}	4.3	22.7
Granite	$10^{-2} - 10^{-5}$	8	13.0
Sandstone		10	9.3–11.1
Synthetic Permafrost:			
(1) Tap Water in Coarse Sand	8.3×10^{-4}	4.5	21.3
(2) Wet Sand, Clay	7.7×10^{-3}	5	17.2
(3) Brown Soil	5×10^{-3}	4.4	20.1
(4) High-Fe Clay	1.5×10^{-2}	7	12.8
(5) High-Al Clay	1.6×10^{-3}	3.3	32.6
(6) Limey Mud	7.2×10^{-3}	3.6	20.3
(7) Silica Mud	7.4×10^{-3}	1	10.9

The basal interface of a glacier or ice sheet is seldom a specular reflector, even when the ice is afloat. In reality, the base of the ice normally contains many irregularities which produce echoes later than that from the nadir point, resulting in a characteristic echo train rather than a single discrete bottom echo. Usually, but not always, the first echo is the largest in amplitude. Figure 118 shows a typical oscilloscope presentation of an echo from the rough base of Skelton Glacier (Jiracek, 1967) [110] – the bottom return can be seen to last approximately 6 μs.

As antennas are moved along the glacier surface, the separate echoes change rapidly in amplitude and position, indicating that roughness on a scale of the order

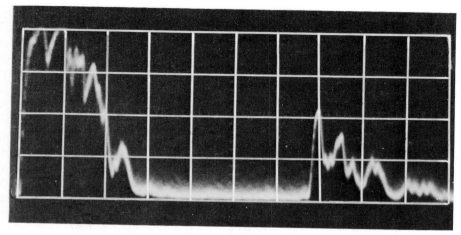

Fig. 118. Oscilloscope A-display recorded on Skelton Glacier, showing a long-lasting bottom echo. The ice thickness is about 1150 m. Vertical scale lines are 2 μs apart. (From Jiracek, 1967.)

of a wavelength (a few meters in ice) is present. This rapid change in the echo characteristic observed while moving along the surface was referred to by A.H. Waite as the 'picket fence' effect. It is this rapidly-changing characteristic of the echo, resulting from interference between arrivals from different reflecting facets of the bottom, that is the basis for the Nye method of ice sheet movement determination described in Section 6.9.

The echo recorded in an aircraft from continuous radar profiling normally shows the same rapid variations in amplitude that are seen for measurements along the surface. The horizontal correlation distance or 'fading length' of the 'picket fence' echo pattern along the flight path is of the order of 20 m, and the echo strength variation is about ±10 dB around the mean value. Harrison (1972) [99] and Berry (1973 [38]; 1975 [39]) have shown that these characteristics can be explained in terms of the statistical properties of the ice–rock interface.

In the following sections we consider the uses that have been made of the characteristics of the bottom reflection to determine the nature of the basal interface.

ANTARCTIC GROUNDED ICE

In his early work in Antarctica, Jiracek (1967) [110] found that, whereas seismic and radio soundings gave approximately the same thickness at South Pole Station, there was a large discrepancy on Roosevelt Island, where the radar method gave thicknesses that were greater by 50–100 m. Since the difference was too large to be attributed to errors in either seismic or radio-wave velocity, both of which were accurately measured, the conclusion seemed inescapable that the radar waves had penetrated into the subglacial material before reflecting. However, although it was clear that the bed must be frozen, careful examination of the problem led to no geologically reasonable permafrost material that would provide an impedance match at the base of the ice and yet not result in such a large loss tangent that the echo from the bottom of a layer 50–100 m thick would be below the detection level. This matter remains a mystery to the present day.

The presence of substantial areas of water beneath the thick Antarctic inland ice was first suggested by Robin *et al.* (1970) [155] to explain an anomalously strong echo from a depth of 4200 m near Sovietskaya. Then, in 1971–72, Oswald and Robin (1973) [135] noted that at a number of sites over the East Antarctic ice sheet (Figure 119) the normal fading characteristics of the bottom echo changed to that which would be expected from an extended smooth surface. At these sites the fading length increased to 200 m or more and the amplitude of echo variations was much reduced. At the same time, the duration of the bottom echo decreased to a length similar to that of a transmitted pulse. There was also a marked increase of 10–20 dB in echo strength when moving from an area of normal fading to one of the smooth areas. These characteristics all indicated that specular reflection was taking place from an extended smooth surface. Futher analysis by Oswald (1975) [134] showed that the observations just mentioned, and also the gentle surface and basal slopes, were explained if the smooth surfaces represented subglacial lakes.

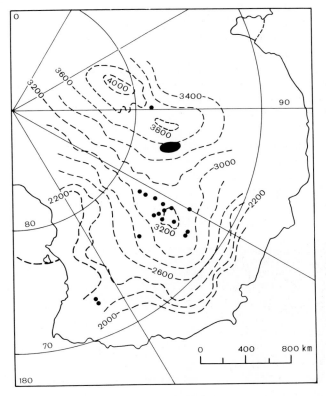

Fig. 119. Map of eastern East Antarctica, showing the locations of subglacial lakes identified by radar sounding. Small areas are shown as black dots; one large lake is also indicated in black. Contour lines show surface elevations. Lakes tend to occur beneath low surface slopes. (From Oswald and Robin, 1973.)

These lakes usually are found in relative low spots of the bedrock topography (Figure 119).

A particularly large subglacial lake (Figure 120) was discovered during the 1974–75 season (Robin *et al.*, 1977) [152]. A smooth oval depression on the ice sheet surface was reported from the same general area near Vostok by the senior aviation navigator with the Soviet Antarctic Expedition in 1959. This large subglacial lake appears to have a surface area of about 50×180 km.

Drewry (1973 [68]; 1976 [71]) has shown how to use normalized variance in autocorrelation functions as a way of determining significant differences in terrain roughness from place to place. The bedrock surface in the East Antarctic subglacial basins generally exhibits statistical properties characterized by only small-scale surface irregularities. The more mountainous flanking areas, however, are distinguished by a substantially rougher bedrock, characterized by high variance and low autocorrelation distances (see Section 6.7). Drewry also carried out a quantitative analysis of the reflection strength in East Antarctica. Making allowance for the dielectric absorption of the ice and geometrical spreading, he calculated reflection coefficients about every 2 km along a 1500 km flight line that crossed the subglacial basins and the neighboring areas. Dividing that track into several sections, he showed that the mean reflection coefficients varied by almost 15 dB between the basins themselves and their flanks. Those differences reflect both the effect of different microscale roughness of the surface, resulting in scattering loss, and change in the specular reflection characteristic resulting from changing dielectric contrasts. Drewry made no attempt to separate these factors.

Detailed measurements on the surface reveal variations that cannot be resolved from flight records. The geophysical survey at Dome C (Bentley *et al.*, 1979 [35]; Shabtaie *et al.*, 1980 [162]) produced radar reflection profiles that show variations in reflection strength of the order of 25 dB over horizontal distances of the order of 100 or 200 m. This characteristic horizontal length is too large to be attributed to normal fading, and is believed by the observers to represent the spacing between subglacial water channels.

The presence of water under the West Antarctic inland ice near the margin of the

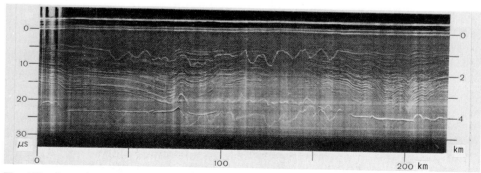

Fig. 120. Intensity-modulated profile of the large suglacial lake shown in Figure 119. The zig-zag flight track crossed the lake twice, between 20 and 70 km, and between 170 and 200 km. (From Robin *et al.*, 1977.)

Ross ice shelf in extensive 'pseudo-ice-shelf' areas (Robin et al., 1970 [155]) was deduced on the basis of elevation and thickness profiles, and was also indicated by variations of echo strength along the flight track. Although these 'pseudo ice shelves' now appear to be parts of the enigmatic ice stream C (Rose, 1979) [158], the interpretation of these regions as having a wet bed is still valid. However, since the basal reflection even from the bed of an ice stream is intermittently weak due to other sources of interference (Rose, 1979) [158], it is unlikely that the short sections of ice with basal grounding suggested in Robin et al. (1970) [155] are in reality grounded at all. This, once again, shows the difficulty in determining the wet or dry nature of the bed from radar echo strengths in the presence of other disturbing factors.

Nevertheless, it now appears that ice streams can be identified definitively by their radar reflecting characteristics because of strong scattered return from creveasses and other near-surface irregularities that are characteristic of active ice streams. Zones of intense surface crevassing that normally mark the boundaries of ice streams will return strong echoes that can be detected across the whole width of the record (Figure 121). Crevassing, much of it probably remnant, also occurs across much of the body of ice streams, resulting in distinct long-tailed bands of hyperbolae that often totally obscure bottom and internal reflections. This 'clutter' resulting from near-surface reflectors can produce a change in tone of the whole reflection record across an ice stream (Figure 121).

In the 1977–78 season, the first large subglacial lake in West Antarctica was detected along the western flank of the Ellsworth Mountains. That water body is at least 13 km in linear extent, occupying a bedrock hollow beneath 3500 m of ice. Several smaller lakes were also identified (Drewry and Meldrum, 1978) [73].

The presence of water was also noted during radar sounding of Lake Stokovoe near Molodezhnaya. The lake boundaries were determined far under the glacier (Figure 122). Drilling results confirmed the sounding data. In 1979–1980 the tractor train traveling on the Mirny-Komsomolskaya–Dome B route was equipped to measure glacier movement and thickness. Radar sounding showed water at the bedrock near Komsomolskaya and Dome B (see Figure 109). Water echoes were obtained on 9 lines 0.2 to 3 km long near Komsomolskaya station.

Two subglacial lakes were found near Dome B, one of them at 76°45′S, 94°30′E,

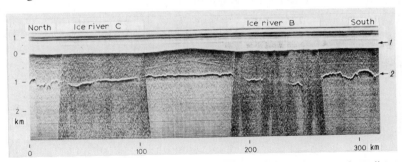

Fig. 121. Intensity-modulated profile across two West Antarctic ice streams, a short distance east of the juncture with the Ross ice shelf. (The actual flight line was from 81.7°S, 139°W to 84.5°S, 141°W). 1 – glacier surface; 2 – glacier bed. (From Rose, 1979.)

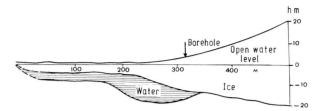

Fig. 122. Profile of subglacial Lake Stokovoe, near Molodezhnaya.

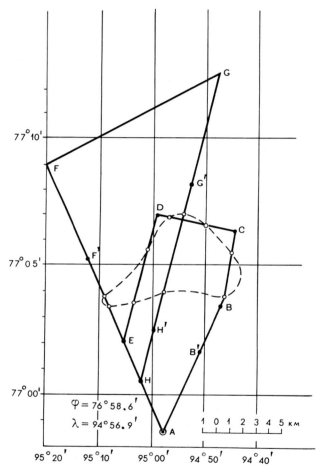

Fig. 123. Map of subglacial lake on the route Mirny–Komsomolskaya–Dome B. Dashed line – deduced outline of the lake. Solid lines – radar-sounding tracks.

and the other at 77°05′S, 94°50′E. The second lake was observed in detail. Five radar lines made over it permitted the determination of its subglacial boundaries (Figure 123). Figure 124 gives the glacier relief along all five lines. Maximum lake depths exceed 500 m, as estimated by interpolation of bedrock slopes. The area of

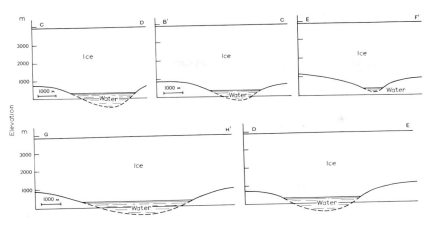

Fig. 124. Subglacial lake sections deduced by interpolation of the subglacial relief along the five radar routes crossings shown in Figure 123.

the lake is approximately 30 km², and the ice thickness above it is 3560 m. The water surface in the lake is under a pressure of 33 MPa – the melting-point at such a pressure is −2.4°C. The peculiarity of the liquid physical state under such conditions is that the entire volume of the lake is under great pressure and at a negative temperature. Theoretical calculations of the temperature field in an ice sheet, made for the Mirny – Komsomolskaya – Dome B region showed that ice melts at the bedrock close to the Dome and 200 km north from Komsomolskaya. The radar showed water echoes just in these locations. In the calculations, heat flow was taken as being equal to $6.8 \times 10^{-2}\,\mathrm{W\,m^{-2}}$. By analogy with Dome C, lakes should also be situated in hollows surrounding Dome B. A detailed radar survey of subglacier relief over a large area is necessary to discover them.

ANTARCTIC ICE SHELVES

Even on the Antarctic ice shelves, where the underlying material is sea water and does not vary from place to place, there are still important variations in the character of the basal interface. Neal (1979) [129] analyzed the apparent roughness of the ice–water interface from airborne survey records gathered over the Ross ice shelf, using the fading patterns in the echo. He found a spectrum of surface roughness, and concluded that the smooth ice–water interface is associated with regions of bottom melting. He then developed a theory showing how a layer of saline ice frozen on to the base of the ice shelf can severely modulate an incident radar wave. Under these circumstances the return from even a smooth ice–water interface will appear rough. Neal then used the patterns of rough and smooth basal ice to map the regions of subglacial melting and freezing (Figure 125). He found that the distribution of these regions appeared to depend mainly on bottom morphology and circulation of sea water beneath the ice.

Along flight tracks within 25 km or so of the seaward margin of the Ross ice shelf,

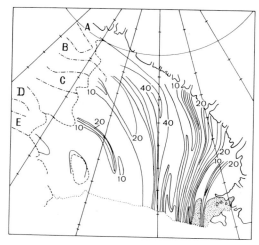

Fig. 125. Map of the Ross ice shelf, showing contours of the reflection coefficient at the ice–water boundary. Shaded regions exhibit a reflection loss >20 dB. The contour interval is 10 dB. (From Neal, 1979.)

very strong reflections appear. Bentley *et al.* (1979) [37] interpreted these as being due to the disappearance, by melting, of the saline ice frozen onto the underside of the shelf, causing poor reflections further upstream.

Similar contrasts were observed during soundings of the Filchner and Ronne ice shelves in 1969–70, although in that case it was the ice of about 1000 m in thickness that showed the strongly-reflecting bottom surface, whereas indications of roughness were found on the thinner parts of the Ronne ice shelf (Robin *et al.*, 1977) [152].

NORTH AMERICAN GLACIERS

An expedition to the Devon Island ice cap in 1973 gave Oswald (1975) [134] an opportunity to make precise and detailed recordings of the echo wave forms. An 8 mm movie camera was used to photograph the A-scope display at predetermined intervals along the sounding track, giving a complete record of the signal envelope accurate to ±1 dB over a range of about 60 dB. A detailed statistical analysis was made of the maximum observed power level in the echo. These analyses led to estimates of the root-mean-square slope in the rough interface, and of the autocorrelation lengths of the bedrock roughness. The estimate of rms slope was derived in two different ways: (1) by the decrease of received power with angle of reflection, and (2) from the observed autocorrelation length. On the basis of this, the bedrock surface was thought to be, in general, gently undulating, containing slopes with an rms value of about 1:40. There is an indication of a small, but significant, proportion of slopes considerably steeper than this value, possibly indicating the presence of small morainal boulders.

The basal echo strengths were also plotted as a function of ice depth (Figure

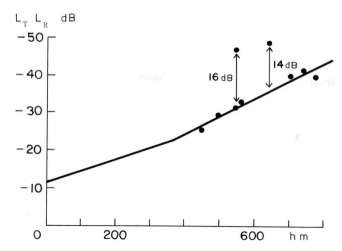

Fig. 126. Echo power vs. ice thickness measured for reflections from the bedrock beneath the Devon Island ice cap (Canada). The echo power shown is the total attenuation relative to that in the case of no absorption or reflection losses. The two straight lines represent the expected absorption according to a model ice cap with a mean temperature of $-20°C$ from the surface to a depth of 400 m, and a mean of $-12°C$ from 400 m to the base at about 800 m. The intercept thus yields the measured reflection loss. (From Oswald, 1975.)

126). This plot revealed two things: (a) after allowance for absorption by the ice, the bedrock reflection coefficient is about -11 dB, which is relatively high; and (b) at two points the reflection was substantially weaker, the difference being highly significant in view of the good fit of the other seven points to a mean line. The poorly-reflecting points were obtained from the western part of the ice cap (see Section 6.6).

In an entirely different type of application of radar sounding Clarke and Goodman (1975) [48] and Goodman et al. (1975) [85] used radar depth measurements and temperature data in drill holes to extrapolate temperatures to the base of Rusty and Trapridge Glaciers in Yukon Territory, Canada. By this means they found an apparent zone of temperate basal ice in the central part of each glacier, although the glaciers have cold beds nearer their edges. This fact may be significant in the surging history of the glaciers.

GREENLAND

One of the early examinations of bedrock roughness came from the extensive analysis of soundings near Camp Century by Robin et al. (1969) [153]. They were, perhaps, the first to calculate roughness of the basal interface in terms of the horizontal fading rate and the length of the returned signal. From these characteristics they estimated numbers approximating the horizontal autocorrelation distance and the main slope of the irregularities of the interface. They found that the angles of the bottom irregularities were surprisingly large, even on a relatively flat inland plain underneath the Greenland ice sheet. Those angles were as great as those commonly observed at shorter wavelengths from aircraft flying over unglacierized terrain. This suggested that the reflection contained a diffuse component produced by irregularities of the same order of size as, or smaller than, the radio wavelength distributed on an otherwise smooth surface, e.g., a boulder-strewn plain.

6.11. Detection of Hidden Crevasses

SURFACE CREVASSES

The need for detection of crevasses was discussed earlier in the book (Section 4.5). For the safety of surface transport operations it is necessary to know their size and location. Air strips built on glacier ice should also be searched for cracks.

Crevasses are known to result from irregular ice flow; their depths reach tens of meters. Snow bridges often top the crevasses. A technique of airborne radar crevasse detection in glacial ice was field tested in the 12th SAE (1967).

The tests showed that, with the aircraft at a height of 300 m, crevasse echoes received at the side lobes of the transmitted radiation pattern (the main lobe being directed to the nadir) could be detected at distances up to approximately 4 km. When surface measurements are made (using an antenna with a broad radiation pattern mounted at a height of 7 m), crevasse echoes are received from a distance of 300 m. These measurements, and the analysis of their results, have demonstrated the possibility of seeing crevasses on the screen of an omni-directional radar with the signal at low grazing angles. However, it is sometimes difficult to identify the echoes and determine crevasse positions due to the spatial resolution of the equipment.

Field trials have shown that the echo is much stronger if antennas are oriented perpendicular to a crevasse rather than parallel to it (See Figure 127). It is, therefore, more effective to use omni-directional, or panoramic, radars for the detection

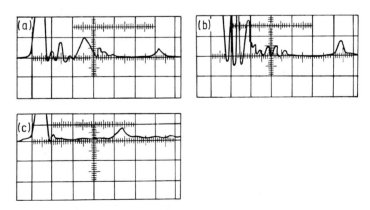

Fig. 127. Radar sounding oscillograms (forward looking radar): a – soundings near Sibiryachka Bay, a crevassed area, antenna oriented parallel to a crevasse; b – the same area, antenna oriented perpendicular to the crevasse; c – sounding where there were no crevasses. The bedrock echo is seen in the right-hand part of each oscillogram.

of crevasses and the determination of their orientations and dimensions. Radars can be fixed either to a surface vehicle or to an aircraft.

For a rough estimate of the total attenuation, a crevasse can be regarded as a flat layer with rough boundaries. The area of the portion of a crevasse irradiated within the space of one resolution element of the radar, S, can be determined using

$$S = \frac{c_i \tau_t}{2} \frac{1}{\cos \alpha} \sqrt{R \frac{c_i \tau_t}{2}},$$

where α denotes the grazing angle relative to the crevasse plane, and R is the distance to the crevasse. G.V. Trepov, in his calculations, made allowance for signal attenuation factors. Crevasse echoes were received from distances more than 100 m away by a radar operating at 440 MHz ($G = 2$, $\tau_t = 0.3\ \mu s$ and 130 dB gain) with antennas 8 m above the ice surface. If the radar frequency were to be 3000 MHz ($G = 300$, $\tau_t = 10$ ns, and 130 dB gain), the crevasse echoes could be received from distances less than 100 m away.

It should be noted here that crevasse detection by radar is strongly hindered by noise, the major source being reflections from surface irregularities and inhomogeneities in the upper ice layer. The noise level depends on the grazing angle, the radiation pattern, the surface roughness, and inhomogeneities in ice density. Using the equipment described in Section 4.5, G.V. Trepov, in the 14th SAE (1969) carried out field trials of crevasse detection by means of an omni-directional with a CI-20 oscilloscope as the signal indicator. Both antennas (transmitting and receiving) were mounted on the roof of a moving laboratory, at a height of 7–8 m above the surface. Observations were carried out at the Molodezhnaya air strip, at a glaciological polygon, and in a heavily crevassed area near Sibiryachka Bay. The oscilloscope record displayed a high level of echoes at distances up to 300 m when the soundings were made in an area with crevasses. No echoes, however, were recorded in an area with monolithic ice (e.g., the air strip).

During the 16th SAE (1971), A.M. Kluga, using high resolution equipment, succeeded in detecting crevasses ranging from 1.5–3 m in width and from 4–12 m in depth. Detection of crevasses covered by snow bridges 0.5–1 m thick is possible only by means of equipment with high spatial resolution.

To estimate the strength of signals reflected from crevasse walls, the effective reflecting area should be known. The echo strength depends on wave polarization, wall roughness, and grazing angle (Figure 128). To measure the width of a crevasse, echoes from its front and rear walls must be distinguished. The front-wall echo is formed by rays similar to the one marked '1' in Figure 128. The beginning of a rear-wall echo is formed by rays similar to those marked '3' in Figure 128. If the selection time interval between the two is larger than the pulse width, the echoes will divide. Maximum time resolution is achieved when the amplitudes of the two echoes are equal. This requires also that the effective reflecting surfaces of the two crevasse walls be equal, a condition that is fulfilled only at certain, optimal sounding angles. Table XXIV lists the values of optimal sounding angles (β_1) and pulse lengths required to detect a crevasse for two different values of its width, a, and the depth of its snow bridge, h.

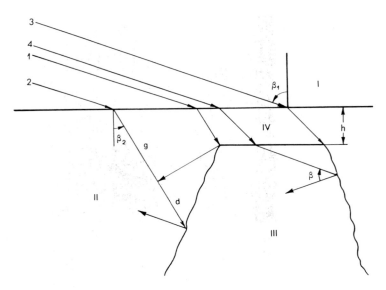

Fig. 128. Diagram of angular relationships in crevasse sounding. 1–4 – ray paths; β_1 and β_2 – sounding angles (angles of incidence); I–IV – air, ice, crevasse, and snow bridge, respectively.

TABLE XXIV

Optimal sounding angles and pulse lengths

	$a = 2\,\text{m}$		$a = 3\,\text{m}$	
	$h = 0.1\,\text{m}$	$h = 1.1\,\text{m}$	$h = 0.1\,\text{m}$	$h = 1.1\,\text{m}$
Crevasse with vertical walls				
β_1	73°	64°	73°	68°
τ_t ns	12	5.3	18.5	11.7
Crevasse with the walls widening downward				
β_1	64°	60°	64°	62°
τ_t ns	11.4	5.0	17.3	11.7
Crevasse with the walls widening upward				
β_1	82°	68°	82°	74°
τ_t ns	12.5	6.1	19.1	12.5

In each experiment the relationship between echoes and β_1 was investigated and recorded at the various frequencies of the equipment used. In one experiment when measurements were made by means of the RS-1, the relationship of the echo to the angle β, the azimuthal angle between the line of sight and the crevasse wall, was studied. When the experiments were completed, the show bridges were destroyed and the remote-sensing and direct measurements were compared. The best agreements were at distances of 1.5 m from a crevasse. An oscillogram of crevasse echoes is shown in Figure 129.

In some cases false echoes were recorded, corresponding to reflections from irregularities on the glacier surface and its boundaries.

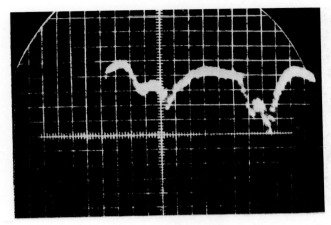

Fig. 129. Oscillogram of a signal reflected from the crevasse wall. The sounder frequency was 440 MHz; the distance from the radar to the crevasse was 15 m; the crevasse width was 3 m; 1 scale division = 5 ns.

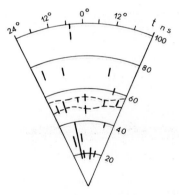

Fig. 130. Panorama of an ice crevasse. Heavy line segments – signals returned by ice surface irregularities and crevasse walls. Dashed lines – real crevasse boundaries.

Figure 130 illustrates a radar sounding picture. It is seen that this method enables a crevasse to be both detected and located (Kluga, 1972 [15]).

In the 18th SAE (1973) a new 400-MHz radar modified for prospective crevasse detection was tested. In this modern equipment an antenna grid, consisting of eight dipoles with a planar reflector, was used. This equipment was designed in the Department of Ice and Ocean Physics in AARI. In 1980, V.I. Poznyak successfully used a centimeter-band, scanning radar for travel security in crevasse-rich areas.

During the 1973–74 season, a short-pulse radar system was evaluated for use as a crevasse detector by Kovacs and Abele (1974) [117]. The output frequency of the system was variable between 1 and 100 MHz, and recording was on a continuous strip chart similar to the systems used for profiling beneath the seabed – in this case, however, the signal was directed forward for crevasse detection. Since the waves propagate fastest through the low-density superficial snow cover, a reflection from

230   CHAPTER 6

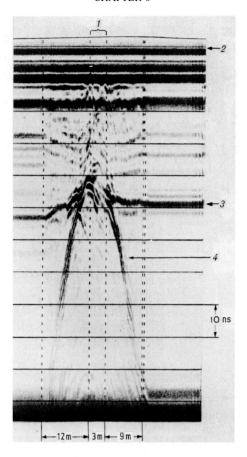

Fig. 131. A graphic record of the reflected radar signal obtained with an impulse radar system when the antenna approached, passed over, and moved beyond a 3-meter-wide snow-bridged crevasse. The record also shows the echo from the well-known brine layer in the McMurdo ice shelf. 1 – crevasse walls; 2 – surface; 3 – brine layer; 4 – reflection from crevasse wall. (From Kovacs and Abele, 1974.)

a crevasse wall near the surface should provide a first return. The field test on the McMurdo ice shelf was successful in showing a 3-m wide crevasse at distances up to 9 m (Figure 131). The system was also tried out in the Pensacola Mountains with the antenna positioned 9 m in front of the tracked vehicle to give additional warning time. The system could be improved by using two antennas to provide better coverage of crevasses being approached at fairly shallow azimuthal angles relative to the travel route.

BURIED SURFACE CREVASSES

As part of the work at the site of the drill hole through the Ross ice shelf at RIGGS station J9 (Clough, et al., 1975 [56]; Jezek, et al., 1979 [108]), a short-pulse radar

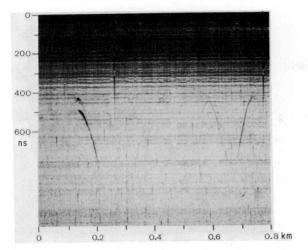

Fig. 132. Impulse radar profile at station J9 showing hyperbolas believed to be diffractions from relict surface crevasses. (From Jezek, 1980.)

reflection survey was carried out (Jezek, 1980) [107]. The system used has a pulse-width of only a few nanoseconds, so it can resolve near-surface details lost in the 250 ns pulse of the 35 MHz radar system. However, the maximum range of the short-pulse radar is only about 100 m.

A sample of the data is shown in Figure 132. The faint hyperbolas apparent throughout the data (particularly strong at a range of 0.15 km) are believed to be diffractions off buried surface crevasses. Symmetrical hyperbolas appear at a nearly constant depth of about 35 m. It is not clear as to which part of the crevasses the symmetrical hyperbolas correspond, but fluctuations in internal layers above the hyperbolas suggest that the diffraction is associated with the lower part of the crevasse. This interpretation is supported by other geophysical evidence at J9 and elsewhere on the Ross ice shelf.

BOTTOM CREVASSES

Bottom crevasses are the inverted equivalent of surface crevasses, found in the basal surface of ice shelves. They are a widespread feature of the Ross ice shelf – evidence for bottom crevasses is present at about half of the 80 or so stations spread over the shelf. They were predicted on the basis of theoretical considerations by Weetman (1973) [190] and first reported from radar soundings by Clough (1974 [53]; 1975 [54]).

A detailed study of a band of bottom crevasses was made at the J9 drill site. The survey comprised a rectangular grid with three 5-km lines 1 km apart, superimposed on a smaller grid of four lines 1.5 km long and 0.25 km apart. Profiling was carried out in directions perpendicular with respect to each other, using SPRI Mark II radar transmitters and receivers operating at 35 and 50 MHz. A good sample of

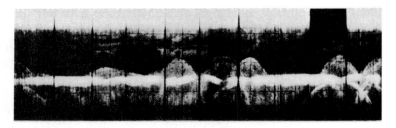

Fig. 133. Radar profile at Ross ice shelf station J9 showing bottom crevasse system. Note that hyperbolas are generated both at the apex and at the base of the bottom crevasses. (From Jezek, 1980.)

the data collected is shown in Figure 133. The numerous hyperbolic features represent reflections from the linear tops of bottom crevasses. Two sets of crevasses are present: one set, represented by the narrow hyperbolas, strikes nearly perpendicular to the traverse path; the other set, yielding much broader hyperbolas, crosses the traverse path at a shallow angle. Jezek et al. (1979) [108] present a careful analysis in which they are able to correlate the crevasses of the two sets, taking into account the fact that the shape of the reflection hyperbola depends upon the angle between the traverse route and the axis of the crevasse.

Along with position and heights, the reflection data yield a measure of the basal width of the bottom crevasses. Just as reflection hyperbolas can originate at the top of the crevasse, hyperbolas can also originate at the intersection of the crevasse with the ice–water boundary (Figure 133). Although it appears, at first, that the latter reflections represent single hyperbolas, more careful examination shows that they are formed of two spatially-separated hyperbolas. The largest crevasses extend as much as 150 m into the ice shelf and have a width at the base of over 100 m. No estimate of the length of the crevasses is possible, except to say that the large crevasses are at least 2.3 km long, since they extend beyond the survey area in both directions.

A study of the bottom crevasse pattern at RIGGS station 'Base Camp' revealed two sets of crevasses. The distribution, orientation, and size of these bottom crevasse patterns must be related to the stresses occurring in the ice shelf, but the stress patterns are too complex to permit any simple analysis.

Jezek (1980) [107] showed evidence from the deflection of internal layers over crevasses at Station J9 that they are probably water-filled. Analyzing the stresses that would occur in a water-filled bottom crevasse and the amount of sinking of the ice shelf surface that would result for an ice shelf thickness of 420 m and a bottom crevasse height of 120 m, he calculated an equilibrium sinking of the surface of 13 m. This is about half the total observed at J9; Jezek attributes the difference to increased deflection resulting from the additional load created by snow drifting into the surface depression. He calculated that the observed amount of sinking has taken place over a period of the order of 100 years.

Substantial additional evidence concerning bottom crevasse fields on the Ross ice shelf comes from airborne radar sounding. Jezek (1980) [107] showed that most of the bottom crevasse fields are associated with grounded ice. He then used the presence of crevasses and their patterns across the ice shelf as clues to the existence of previously unidentified ice rises within the ice shelf.

The crevasses in all groups that Jezek (1980) [107] has observed initially appear as bright diffraction hyperbolas. In most cases they are no longer detectable after about 100 km of downstream flow (taking about 250 years). The disappearance of crevasses can be shown, by a simple calculation, to be too rapid to be attributable to simple vertical strain. Jezek's interpretation is that they have become invisible to electromagnetic waves by freezing shut. Since an ice–water boundary has a reflection coefficient 40 dB higher than a boundary between freshwater ice and sea ice, freezing of the crevasse would reduce returned amplitudes by 40 dB. Jezek (1980) [107] also presented careful analyses of the diffraction caused by the top and bottom corners of a bottom wedge. For a frozen bottom crevasse whose apex is 300 below the surface and whose height and width are both 120 m, the relative amplitude of the diffracted arrival off the apex is almost 60 dB below that of a perfect reflector located 420 m below the receiver. Many other useful diffraction analyses were also presented in Jezek's work.

6.12. Estimation of Glacial Water Content

Estimation of the water mass stored in glaciers is an important problem of hydrology. Glaciers (Antarctic and Greenland ice caps, large Arctic glaciers), being unique accumulators of fresh water, are natural reserves for the future of our planet and mankind. Mountain glaciers of the Caucasus and Central Asia, situated in arid regions of the U.S.S.R., supply water in those seasons when it is most needed. This important water source promotes the development of irrigation farming on a more efficient basis. The need for the estimation of glacial water content is obvious.

To assess the water mass stored in glaciers, the ice mass should be known, i.e., the volume of the ice and its variations in density. At the present stage, this complicated task can be solved with sufficient accuracy. To calculate the mass of a glacier, models can be used that realistically simulate glacier structure.

The remote-sensing radar technique for measuring ice thickness that has been developed and put into practice in the Soviet Union not only enables ice thickness to be measured but also permits the inner structure (inhomogeneity) and state (temperature) of glaciers and their speed of movement to be studied. In the 1960s experiments were carried out to study variations in ice density that depend on the value of the static pressure.

Previously, the volume of a glacier was determined as a product of its area and its mean thickness. Until recently data on glacier thickness were limited to spot values. It is obvious that sparsity of such important data would prevent accurate quantitative determination of volume, hence also of mass. That is why the limited information on ice volume that is presently available is approximate. Thus, the volume of glaciers on Novaya Zemlya was estimated by Govorukha and others (1974) [14] to be 4500 km^3, while Bogorodsky and others (1974) [12] indicated it to be 5250 km^3, and a few years ago the volume of Novaya Zemlya glaciers was found to be 3500 km^3. It is apparent that radar sounding can become a reliable method giving sufficiently reasonable estimates of glacier mass and its water equivalent (Bogorodsky, 1978) [5].

GENERAL CONSIDERATIONS

The thickness of glaciers measured by a radar is known either at individual points or along a complete section along the flight line (or several such sections). The flight direction cannot be chosen correctly in all areas. Hence the available information on the thickness of most glaciers is far from complete. It is thus desirable to make use of those methods that enable an approximate estimation of glacial mass to be made on the basis of measurements obtained along a single section. The proposed approximate method involves the construction of a theoretical glacier. This requires some additional information about the glacier, i.e., the portions of the whole glacier area that lies on each side of the section. Also, changes in density along the section should be considered. Ice density is known to depend on pressure, which in turn is determined by the glacier thickness. In the future, as more data on each glacier become available, calculation methods will be developed based on measurements of glacier thickness along a special grid.

MASS DETERMINATION BY A PROFILE MEASUREMENTS

Radar techniques provide us with the positions of upper and lower (bedrock) glacier boundaries, which determine its thickness. Let 2ℓ be the section length (Figure 134) – the x-axis is assumed to coincide with the direction of movement of

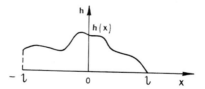

Fig. 134. Hypothetical glacier profile along a single radar-sounding track.

the radar platform (aircraft). The origin of the coordinates lies in the middle of the section. $h(x)$ is the glacier thickness. x is normally measured in kilometers and $h(x)$ in meters. $\rho = \rho(h)$ is the (vertically-averaged) ice density, a function of the glacier thickness. The mass along the section can then be expressed by

$$M_0 = \int_{-\ell}^{\ell} dx \int_{0}^{h(x)} \rho(h)\, dh. \tag{6.16}$$

The $\rho(h)$ relationship was found experimentally (Bogorodsky and Fedorov, 1967) [11]. Bogorodsky has indicated that the $\rho(h)$ function in glaciers with thicknesses up to 1100 m is described with an accuracy of about 10^{-3} by

$$\rho(h) = \rho_0 + \rho_1 h;$$

where $\rho_0 = 0.900\,\mathrm{mg\,m^{-3}}$ and $\rho_1 = 3.884 \times 10^{-5}\,\mathrm{mg\,m^{-4}}$. Substituting into (6.16) we get

$$M_0 = \int_{-\ell}^{\ell} h(x)\left[\rho_0 + \frac{\rho_1}{2} h(x)\right] dx. \tag{6.17}$$

Approximation of h(x) function by a parabola

In this section, and further on in the text, a technique to estimate glacier water content when the glacier thickness is measured only along one section is described. The first step involves the approximation of $h(x)$ by a parabola, assuming that the mass for the available profile is equal to that for a parabolic one. Let us consider the parabola crossing the axis at points $(-\ell)$ and ℓ:

$$h = (\ell^2 - x^2)\frac{H_0}{\ell^2}, \tag{6.18}$$

where H_0 denotes the largest value (at $x = 0$). Substituting (6.18) into (6.17), we obtain the formula to calculate the water content along the parabolic profile:

$$M(H_0) = 2\int_0^\ell \frac{H_0}{\ell^2}(\ell^2 - x^2)\left[\rho_0 + \frac{1}{2}\rho_1\frac{H_0}{\ell^2}(\ell^2 - x^2)\right]dx$$

$$= 2\rho_0\frac{H_0}{\ell^2}\int_0^\ell (\ell^2 - x^2)dx + \rho_1\frac{H_0^2}{\ell^4}\int_0^\ell (\ell^2 - x^2)dx \tag{6.19}$$

$$= 2\ell H_0\left(\frac{2}{3}\rho_0 + \frac{4}{15}\rho_1 H_0\right).$$

Let us determine H_0 from the condition that $M(H_0)$ is equal to M_0:

$$\frac{8}{15}\rho_1\ell H_0^2 + \frac{4}{3}\rho_0\ell H_0 = M_0;$$

then

$$H_0 = \frac{15}{4}\frac{\sqrt{\frac{1}{9}\rho_0^2 + \frac{2}{15}\frac{M_0\rho_1}{\ell}} - \frac{1}{3}\rho_0}{\rho_1}. \tag{6.20}$$

It should be noted here that there can be a number of profiles with equal water content. Let us consider a profile 2ℓ long, composed of two parabolas (Figure 135):

$$h(x) = \begin{cases} \dfrac{H_0}{x_0^2}x(2x_0 - x), & x \leq x_0 \\ \dfrac{H_0}{(2\ell - x_0)^2}(2\ell - x)(x - 2x_0 + 2\ell), & x \geq x_0. \end{cases}$$

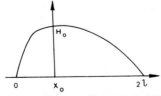

Fig. 135. An example of the family of parabolic profiles, all of which represent the same water content (volume).

The position of x_0 is arbitrary and the value of H_0 is given. The plot of $h(x)$ is presented in Figure 135. Then the water content of a glacier can be calculated making use of the single parabolic profile for $(0, x_0)$ portion of the curve; it is

$$\frac{H_0}{x_0^2}\int_0^{x_0} x(2x_0 - x)\left[\rho_0 + \frac{\rho_1 H_0}{2x_0^2}(2x_0 - x)\right]dx$$

$$= H_0 x_0 \left[\frac{2}{3}\rho_0 + \frac{4}{15}\rho_1 H_0\right]$$

and, for $(x_0, 2\ell)$,

$$H_0(2\ell - x_0)\left[\frac{2}{3}\rho_0 + \frac{4}{15}\rho_1 H_0\right].$$

Thus, the water content along the whole chosen parabolic profile is

$$2\ell H_0\left(\frac{2}{3}\rho_0 + \frac{4}{15}\rho_1 H_0\right). \tag{6.21}$$

Equation (6.21) shows the water content to be independent of the shape of the profile. Comparing (6.19) and (6.21), we can clearly see the possibility of substituting for the real profile another one consisting of two parabolas in which the the H_0 value will be calculated using (6.20). The water content of a glacier determined by means of such a profile is equal to that calculated by a given real profile.

'THEORETICAL GLACIER'

Let S_1 and S_2 be the portions of the whole glacier area that lie on the two sides of the profile (Figure 136(a)). Instead of these portions, half-ellipses with the same areas will be used. ℓ thus determines one of the semi-axes of the ellipse. The other semi-axes are $a_1 = \dfrac{2S_1}{\pi\ell}$; $a_2 = \dfrac{2S_2}{\pi\ell}$. Let us express the glacier profile along the line

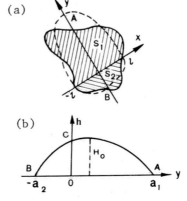

Fig. 136. A theoretical glacier (a) and a profile along line AB (b).

AB (the y-axis, see Figure 136(b)), by a parabola crossing the axes at points A (0, a_1), B (0, a_2), and C (0, H_0).

The parabola is defined by the expression

$$h = \frac{H_0}{a_1 a_2}(a_1 - y)(y + a_2) = H_0(y). \tag{6.22}$$

A 'theoretical glacier' can be constructed in the following way (Figure 137(a)). At each point on the line AB, a cross-profile is constructed, parallel to the x axis. The glacier thickness along these cross-profiles is expressed by the parabola

$$h = [\ell^2(y) - x^2]\frac{H_0(y)}{\ell^2(y)},$$

where $H_0(y)$ is calculated by means of (6.22), while $\ell(y)$ is determined by the equation for the appropriate ellipse.

The water content of such a glacier is then determined. For this, a portion of the glacier, e.g., where $y \geq 0$, is considered.

The equation for an ellipse in parametric form (Figure 137(b)) is

$$x = \ell \sin \varphi; \qquad y = a_1 \cos \varphi.$$

The semi-axis of the ellipse parallel to the x-axis, which is a function of φ, is given by $\ell(y) = \ell(\varphi) = \ell \sin \varphi_0$. Maximum glacier thickness $H_0(y)$ along this profile will be, according to (6.22),

$$H_0(\varphi) = \frac{H_0}{a_2}(1 - \cos \varphi)(a_2 + a_1 \cos \varphi).$$

Then the mass $M(\varphi)$ of the glacier along this profile will be, according to (6.19),

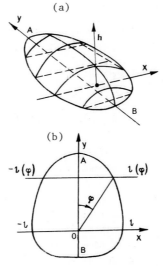

Fig. 137. Diagram of a theoretical glacier: a – perspective of the surface, b – glacier outline.

$$M(\varphi) = 2\ell(\varphi)H_0(\varphi)\left[\frac{2}{3}\rho_0 + \frac{4}{15}\rho_1 H_0(\varphi)\right]$$

$$= \frac{2H_0}{a_2}\ell \sin\varphi(1 - \cos\varphi)(a_2 + a_1\cos\varphi)$$

$$\left[\frac{2}{3}\rho_0 + \frac{4}{15}\rho_1\frac{H_0}{a_2}(1 - \cos\varphi)(a_2 + a_1\cos\varphi)\right],$$

and the mass of the portion of the glacier defined by ($y \geq 0$) will be

$$M_1 = \int_0^{\pi/2} M(\varphi)\sin\varphi\, d\varphi$$

$$= \frac{2H_0\ell a_1}{a_2}\left[\frac{2}{3}\rho_0\int_0^{\pi/2}\sin^2\varphi(1 - \cos\varphi)(a_2 + a_1\cos\varphi)\, d\varphi \right.$$

$$\left. + \frac{4}{15}\rho_1\frac{H_0}{a_2}\int_0^{\pi/2}\sin^2\varphi(1 - \cos\varphi)^2(a_2 + a_1\cos\varphi)^2\, d\varphi\right].$$

Carrying out the integrations, and letting $a_1/a_2 \equiv \mu$, the following formula is obtained:

$$M_1 = \frac{4}{3}H_0\ell\rho_0 a_1\left[\frac{1}{3}(\mu - 1) + \frac{\pi}{16}(4 - \mu)\right] +$$

$$+ \frac{8}{15}H_0^2\rho_1\ell a_1\left[\frac{2}{3}(1 - \mu)\left(\frac{2}{5}\mu - 1\right) + \right.$$

$$\left. + \frac{\pi}{4}\left(1 + \frac{1 - 4\mu + \mu^2}{4} + \frac{1}{8}\mu^2\right)\right]. \tag{6.23}$$

The first term in (6.23) determines the mass of the glacier for ice density ρ_0; the second is a correction for the density increase with depth. To get the mass of the second portion of the glacier, i.e., for $y \leq 0$, a_2 should be substituted for a_1 in (6.23), and μ should be redefined by $\mu \equiv \frac{a_2}{a_1}$.

On the types of glaciers for which water content can be accurately calculated.

The glacier should be divided into two ellipses by a radar profile. Consider sections parallel to that profile. The upper boundary of the section is assumed to consist of two parabolas with maximum height $H_0(y)$, determined by (6.22). The water content, as shown above, is independent of the position of the point x_0 where the maximum thickness is reached. Thus, the position of point x_0 can vary from place to place. The results of calculations will correspond to a 'theoretical glacier'. (See Figure 137(a)). The value of integral M will not change if the sections are laterally displaced relative to each other. In this way, accurate results are obtained for glaciers that, in plan, do not have the form of ellipses (Figure 138).

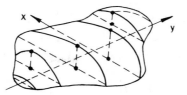

Fig. 138. A glacier which does not have the shape of an ellipse in plan; its water content still can be estimated accurately.

CALCULATION OF GLACIAL MASS

If there are several profiles along which ice thickness was measured, the calculations can be made according to (6.23) for each profile separately. The results obtained are averaged. The physical dimensions of the quantity M are determined from (6.23); a is measured in kilometers (1000 m), ℓ is measured in kilometers (1000 m), H_0 is measured in meters (6.20), ρ_0 in mg m^{-3}, and ρ_1 in mg m^{-4}. Then

$$M \simeq \text{m} \times 1000\,\text{m}\,\text{mg}\,\text{m}^{-3}\,1000\,\text{m} + \text{m}^2\,\text{mg}\,\text{m}^{-4}\,1000\,\text{m} \times 1000\,\text{m}$$
$$\simeq 10^6\,\text{mg}$$

Thus, masses obtained from this formula are in millions of tons.

The input data for the computer program are the identifying number of the profile; the number of points on the profile, N; the distance between thickness measurement points, Δx; S_1 and S_2, the areas of the two portions of the glacier (in km^2); a set of N values of $h(x)$; and ice thicknesses in x_k points. The calculation algorithm involves the determination of the value, M_0, using the formula for rectangles:

$$M_0 = \Delta x \sum_{k=1}^{N} h(x_k)\left[\rho_0 + \frac{\rho_1}{2}h(x_k)\right].$$

Then, using (6.20) and (6.23), H_0, M_1, and M_2 are calculated. The sum of $M_1 + M_2$ is the glacial mass sought.

The profile number, H_0, M_1, M_2, and the total mass are printed out. The program is written in ALGOL; computations are made by a M-222 computer with an ALFA compiler.

How to get averages from several profiles

Denote the water mass of a glacier by M_k ($k = 1, 2, 3 \ldots n$), estimated by profiles L_k, the profile length being ℓ_k. To average these values, an assumption is made: the longer the profile, the more reliable the value obtained is. Accordingly, letting P_k be $\ell_k/\Sigma_{k=1}^{n}\ell_k$, so that $\Sigma_{k=1}^{n}P_k = 1$, the glacial water mass is calculated from the expression

$$M = \sum_{k=1}^{n} P_k M_k.$$

Calculations of the water content of the Severnaya Zemlya glaciers by means of radar measurements.

Table XXV shows water masses of Severnaya Zemlya glaciers estimated by the profile method. Radar sounding of ice thickness on Akademia Nauk, Schmidt, Albanov, Pioner, and Ushakov Glaciers was carried out in September–October 1968.

The water content of Vavilov Glacier was calculated using several radar profiles. It follows from the above that a series of flight lines would give a more accurate ice volume (and hence mass) if the distances between the flight lines, and between points along a flight line, were not very large. Maximum reliability of the estimates is possible when the number of points along a flight line has been chosen according to Kotelnikov's theorem.

In 1975 a large amount of information was obtained which can be used to calculate the volume of Vavilov Glacier with a better accuracy than that obtained by using the methods described in previous sections. To measure ice thickness on Vavilov Glacier accurately, a special measuring scheme was designed that made use of both surface and airborne radar soundings. The glacier surface was divided into squares of area $1.072 \times 4 \text{ km}^2$, the coefficient 1.072 being the result of a more precise determination of glacier area on a map of a larger scale than the one that was used initially. The basic square was chosen so that its side was of the same length as the distance between flight lines in the central and northern portions of the glacier, which was divided into squares by north–south and east–west lines. Interpolation between thickness profiles was made by assuming that changes in the

TABLE XXV

Estimates of the masses of Severnaya Zemlya glaciers (by a single profile and by several profiles)

Glacier	Flight line	ℓ (km)	P_k	Water mass along profile (million tons)	Resulting water mass (million tons)
Akademii nauk	D	43.5	1.000	2617186.9	2617186.9
Schmidt	L	21.0	0.477	49586.8	57076.5
	M	23.0	0.523	63907.4	
Albanov	Z	16.5	0.465	57933.9	57795.4
	I	19.0	0.535	57675.0	
Pioner	A	11.5	0.295	42675.1	44132.0
	B	14.25	0.365	57342.7	
	V	13.25	0.340	31213.8	
Ushakov	P	22.0	0.537	26043.9	36022.3
	T	19.0	0.463	47595.5	
Vavilov	—	—	—	—	520000.0

Note: For positions of the flight lines, see Bogorodsky and Fedorov, 1970 [12].

bedrock profile between flight lines were gradual (graphical averaging). The changes in bedrock topography were determined from cross-sections, and from adjacent flight lines; at the eastern and south-eastern glacial margins, the ice thickness was extrapolated. In the vicinity of the radar measurements, the ice thickness was assumed to be the same as measured. A correction was introduced for an increase in ice thickness in the northern and central portions of the glacier. At large distances from flight lines the following assumptions were made: the bedrock relief in the central part of the glacier is flat and uniform, the surface of the glacier becomes lower close to its boundaries, this elevation decrease obeys the parabolic law (while glacier thickness near the boundaries (3–5 km) changes linearly – this assumption was confirmed by radar measurements), the ice volume in a column is the product of its mean thickness and its area, and at the glacier's edge it is determined by part of a square.

The volume of Vavilov Glacier was calculated, using the foregoing assumptions, to be 579.8 km^3, its area being 1816.8 km^2. Taking 0.9 mg m^{-3} as the mean ice density, the mass of Vavilov Glacier was calculated to be $M = 520\,000$ million tons. It is seen in the Table that the mass of the six named glaciers is approximately 2.8×10^{12} tons. Radar measurements on other glaciers of the archipelago will make it possible to determine their masses more accurately.

The described method can be used to estimate the mass of any glacier, including mountain glaciers. However, radar measurements of mountain glaciers require radars with different specifications than those used at Severnaya Zemlya.

References: Chapter 6

[1] *Atlas of Antarctica*, Vol. I, Leningrad, Main Department of Cartography of the U.S.S.R., 1966, 66 pp.

[2] Budd, W.F. *Ice mass dynamics*. Translated from the English, Leningrad, Hydrometeoizdat, 1975, 236 pp.

[3] Belousova, I.M. Bogorodsky, V.V., Danilov, O.V., Ivanov, I.P. 'Glacier movement investigations by means of laser'. *Reports of the Academy of Sciences of the U.S.S.R.*, 1971, **1999** (5) 1055–1057.

[4] Bogorodsky, V.V. *Physical methods of glacier investigations*. Leningrad, Hydrometeoizdat, 1962, 212 pp.

[5] Bogorodsky, V.V. 'Complex method of water mass estimation in glaciers'. *Meteorology and hydrology*, 1978, No. 11, 70–77.

[6] Bogorodsky, V.V., Trepov, G.V., Fedorov, B.A. 'Possibility of laser use for ice cover dynamic investigations'. *Trans. AARI*, 1970, **295**, 32–34.

[7] Bogorodsky, V.V., Trepov, G.V., Fedorov, B.A. 'Focusing factor in radar sounding of media. *Trans. AARI*, 1974, 324, 4–11.

[8] Bogorodsky, V.V., Trepov, G.V. Fedorov, B.A. 'Amplitude fluctuations of radar signals in glacier sounding'. *Trans. AARI*, 1974, **324**, 20–27.

[9] Bogorodsky, V.V., Trepov, G.V., Fedorov, B.A. 'Polarization variations of radar signals in sounding of glaciers in the Arctic'. *J. Technical Physics*, 1976, **46** (2) 366–372.

[10] Bogorodsky, V.V., Trepov, G.V., Sheremetyev, A.N. 'Radar sounding measurements for determining the thickness and velocity of the Antarctic ice sheet'. *Trans. Academy of Sciences of the U.S.S.R. Physics of the Earth*, 1979, No. 1, 95–100.

[11] Bogorodsky, V.V., Fedorov, B.A. 'Radar sounding of glaciers'. *J. Technical Physics*, 1967, **37** (4) 782–788.

[12] Bogorodsky, V.V., Fedorov, B.A. 'Radar sounding of glaciers of Severnaya Zemlya'. *Trans.*

AARI, 1970, **295**, 5–16.

[13] Boyarsky, V.I., Fedorov, B.A. 'A few peculiarities of radar signals in sounding of glaciers of Severnaya Zemlya'. *Trans. AARI*, 1978, **359**, 56–60.

[14] Govorukha, L.S., Ivanov, V.V., Chizhov, O.P. 'Glacial water resources and ice flow in the Arctic'. *Problems of Arctic and Antarctic*, 1974, No. 45, 5–12.

[15] Kluga, A.M. 'Energy characteristics of radar signals from glacial crevasses'. *Trans. RICA*, 1972, **228**, 56–60.

[16] *Ice and Snow*. Edited by Kingery, W.D. (Translated from the English and edited by Saveljev, B.A.). Peace, 1966, 478 pp.

[17] Macheret, Yu.Ya., Gromyko, A.N., Juravlev, A.V. 'Radar sounding of Spitsbergen glaciers'. *Materials of glaciological investigations*, 1980, No. 38, 279–287.

[18] Macheret, Yu.Ya., Juravlev, A.V., 'Interpretation of results of radar sounding of Spitsbergen glaciers from a helicopter'. *Materials of glaciological investigations*, 1973, No. 22, 109–131.

[19] *World Water Balance and Water Resources of the Earth*. Leningrad, Hydrometeoizdat, 1974, 638 pp.

[20] Pasynkov V.V. *Materials for Electronic Engineering*. Moscow, High School, 1980, 406 pp.

[21] Pustylnick E.T. *Statistical Methods of Analysis and Processing of Data*. Moscow Science, 1968, 196 pp.

[22] Trepov, G.V. 'Estimation of glacial temperature profiles from radar sounding data'. *Bull. SAE*, 1970, No. 79, 53–55.

[23] Fedorov, B.A. 'Measurements of the mean velocity of vertical propagation of pulse signals in glaciers of Severnaya Zemlya'. *Trans. AARI*, 1978, **359**, 48–55.

[24] Ackley, S.F., Itagaki, K. 'Strain effect on the dielectric properties of ice'. U.S. Army Terrestrial Sci. Center, Tech. Note, 1969.

[25] Ackley, S.F., Keliher, T.E. 'Ice sheet internal radio-echo reflections and associated physical property changes with depth'. *J. Geophys. Res.*, 1979, **84** (B10) 5675–5680.

[26] Allison, I., Frew, R., Knight, I. 'Bedrock and surface topography of the coastal regions of Antarctica between 48°E and 64°E'. *Polar Record*, 1982, **21** (132) 241–252.

[27] Ambach, W., Denoth, A. 'Studies on the dielectric properties of snow'. *Zeitschrift für Gletscherkunde und Glazialgeol.*, 1972, **8** (1–2) 113–123.

[28] Annan, A.O. 'Radio interferometry depth sounding. Part 1: Theoretical discussion'. *Geophysics*, 1973, **38**, (3) 557–580.

[29] Bailey, J.T., Evans, S. 'Radio echo sounding on the Brunt ice shelf and in Coats Land, 1965'. *British Antarctic Survey Bull.*, 1968, No. 17, 1–12.

[30] Banos, A. *Dipole radiation in the Presence of a Conducting Half-Space*. New York, Pergamon Press, 1966.

[31] Barnard, H.P. 'Radio echo sounding in western Dronning Maud Land, 1974'. *South African J. Antarctic Res.*, 1975, No. 5, 37–41.

[32] Beitzel, J.E. 'Geophysical exploration in Queen Maud Land, Antarctica'. In: A.P. Crary, ed, *Antarctic Snow and Ice Studies II*, Amer. Geophys. Union, Antarctic Res. Ser., 1971, Vol. 16, pp. 39–87.

[33] Bentley, C.R. 'Seismic anisotropy in the West Antarctic ice sheet'. In: A.P. Crary, ed., *Antarctic Snow and Ice Studies II*, Amer. Geophys. Union, Antarctic Res. Ser., 1971, Vol. 16, pp. 131–177.

[34] Bentley, C.R. 'Advances in geophysical exploration of ice sheets and glaciers'. *J. Glaciol.*, 1975, **15** (73) 113–134.

[35] Bentley, C.R. *et al.* 'Geophysical investigation of the Dome C area'. *Antarctic J. U.S.*, 1979, **16** (5) 98–100.

[36] Bentley, C.R. 'Variations in valley glacier activity in the Transantarctic Mountains as indicated by associated flow bands in the Ross ice shelf'. – IAHS Pub. No. 131 (*Sea Level, Ice and Climatic Change: Proc. Canberra Symposium*, December 1979), 1981, pp. 247–251.

[37] Bentley, C.R., Clough, J.W., Jezek, K.C., Shabtaie, S. 'Ice thickness patterns and the dynamics of the Ross ice shelf'. *J. Glaciol.*, 1979, **24** (90) 287–294.

[38] Berry, M.V. 'The statistical properties of echoes diffracted from rough surfaces'. *Phil. Trans. Roy. Soc. London*, 1973, Ser. A, **273** (1237) 611–654.

[39] Berry, M.V. 'The theory of radio echoes from glacier beds'. *J. Glaciol.*, 1975, **15** (73) 65–74.

[40] Blankenship, D.D. et al. Seismic and magnetic indications of subglacial geology, Dome C, Antarctica' (abstract). Paper presented at AGU Midwest Meeting, DeKalb, 18–19 September 1980.
[41] Bourgoin, J.-P. 'Quelques caracteres analytiques de la surface et du socle de l'inlandsis groenlandais'. *Annales de Geophysique*, 1956, **12** (1) 75–83.
[42] Budd, W.F. 'Stress variations with ice flow over undulations'. *J. Glaciol.*, 1971, **10** (59) 29–48.
[43] Budd, W.F., Carter, D.B. 'An analysis of the relation between the surface and bedrock profiles of ice caps'. *J. Glaciol.*, 1971, **10** (59) 197–209.
[44] Budd, W.F., Young, N.W. 'Results from the IAGP flowline study inland of Casey, Wilkes Land, Antarctica'. *J. Glaciol.*, 1979, **24** (90) 89–101.
[45] Calkin, P.E. 'Radio echo sounding records from southern Victoria Land'. *Antarctic J. U.S.*, 1971, **6** (5) 208–209.
[46] Calkin, P.E. 'Subglacial geomorphology surrounding the ice-free valleys of southern Victoria Land, Antarctica'. *J. Glaciol.*, 1974, **13** (69) 415–430.
[47] Camplin, G.C., Glen, J.W. 'The dielectric properties of HP-doped single crystals of ice'. In: Whalley, E., Jones, S.J., Gold, L.W., eds., *Physics and Chemistry of Ice: Papers presented at the Symposium on the Physics and Chemistry of Ice*, Ottawa, Canada, 14–18 August 1972, Royal Society of Canada, Ottawa, 1973, pp. 256–261.
[48] Clarke, G.K.C., Goodman, R.H. 'Radio-echo sounding and ice-temperature measurements in a surge-type glacier'. *J. Glaciol.*, 1975, **14** (70) 71–78.
[49] Clough, J.W., Bentley, C.R. 'Electromagnetic sounding of glacial and shelf ice'. *Antarctic J. U.S.*, 1967, **2** (4) 119–120.
[50] Clough, J.W., Bentley, C.R. 'Measurement of electromagnetic wave velocity in the East Antarctic ice sheet'. In: Gow, A.J. et al., eds., *Proc. ISAGE Symposium*, IASH Pub., No. 86, 1970, pp. 115–128.
[51] Clough, J.W. 'Radio echo sounding: brine percolation layer'. *J. Glaciol.*, 1973, **12** (64) 141–143.
[52] Clough, J.W. 'Propagation of radio waves in the Antarctic ice sheet'. Ph. D. Thesis, Geophys. and Polar Res. Center, Univ of Wisconsin-Madison, 1974, 119 pp.
[53] Clough, J.W. 'RISP radio echo soundings'. *Antarctic J. U.S.*, 1974, **9** (4) 159 pp.
[54] Clough, J.W. 'Bottom crevasses in the Ross ice shelf'. *J. Glaciol.*, 1975, **15** (73) 457–458.
[55] Clough, J.W. 'Radio echo sounding: reflections from internal layers in ice sheets'. *J. Glaciol.*, 1977, **18** (78) 3–14.
[56] Clough, J.W., Jezek, K.C., Robertson, J.D. 'RISP drill site survey'. *Antarctic J. U.S.*, 1975, **10** (4) p. 148.
[57] Collins, I.F. 'On the use of the equilibrium equations and flow law in relating the surface and bed topography of glaciers and ice sheets'. *J. Glaciol.*, 1968, **7** (50) 199–204.
[58] Cook, J.C. 'RF electrical properties of salty ice and frozen earth'. *J. Geophys. Res.*, 1960, **65** (6) 1767–1771.
[59] Crary, J.H., Crombie, D.D. 'Antarctic ice cap attenuation rates of VLF signals determined from short and long great circle paths'. *Radio Science*, 1972, **7** (2) 233–238.
[60] Crary, A.P., Robinson, E.S., Bennett, H.F., Boyd, W.W. 'Glaciological studies of the Ross ice shelf, Antarctica, 1957–60'. IGY Glaciol. Report 6, Am. Geog. Soc., 1962, 193 pp.
[61] Cumming, W.A. 'The dielectric properties of ice and snow at 3.2 centimeters'. *J. Appl. Phys.*, 1952, **23** (7) 768–773.
[62] Davis, J.L., Halliday, J.S., Miller, K.J. 'Radio echo sounding on a valley glacier in East Greenland'. *J. Glaciol.*, 1973, **12** (64) 87–91.
[63] Doake, C.S.M. 'Glacier sliding measured by a radio echo technique'. *J. Glaciol.*, 1975, **15** (73) 89–93.
[64] Doake, C.S.M. 'Bottom sliding of a glacier measured from the surface'. *Nature*, 1975, **257** 780–782.
[65] Doake, C.S.M., Gorman, M., Paterson, W.S.B. 'A further comparison of glacier velocities measured by radio-echo survey methods'. *J. Glaciol.*, 1976, **17** (75) 35–38.
[66] Drewry, D.J. 'Subglacial morphology between the Transantarctic Mountains and the South Pole'. In: R.J. Adie, ed., *Antarctic Geology and Geophysics*, Olso, Universitetsforlaget, 1971, pp. 693–703.

[67] Drewry, D.J. 'The contribution of radio echo sounding to the investigation of Cenozoic tectonics and glaciation in Antarctica'. Inst. British Geographers, Special Pub., 1972, No. 4, pp. 43–57.
[68] Drewry, D.J. 'Sub-ice relief and geology of East Antarctica'. Ph. D. Thesis, University of Cambridge, 1973, 217 pp.
[69] Drewry, D.J. 'Comparison of electromagnetic and seismic-gravity ice thickness measurements in East Antarctica'. *J. Glaciol.*, 1975, **15** (73) 137–150.
[70] Drewry, D.J. 'Terrain units in eastern Antarctica'. *Nature*, 1975, **256**, 194–195.
[71] Drewry, D.J. 'Sedimentary basins of the East Antarctic craton from geophysical evidence'. *Tectonophysics*, 1976, **36** (1–3) 301–314.
[72] Drewry, D.J. 'Ice flow, bedrock, and geothermal studies from radio-echo sounding inland of McMurdo Sound, Antarctica'. In: C. Craddock, ed., *Antarctic Geoscience*, University of Wisconsin Press, 1982, pp. 977–983.
[73] Drewry, D.J., Meldrum, D.T. 'Antarctic airborne radio echo sounding, 1977–78'. *Polar Record*, 1978, **19** (120) 267–273.
[74] Drewry, D.J., Meldrum, D.T., Jankowski, E. 'Radio echo and magnetic sounding of the Antarctic ice sheet, 1978–79'. *Polar Record*, 1980, **20** (124) 43–57.
[75] Evans, S. 'Dielectric properties of ice and snow – a review'. *J. Glaciol.*, 1965, **5** (42) 773–792.
[76] Evans, S. 'Progress report on radio echo sounding'. *Polar Record*, 1967, **13** (85) 413–420.
[77] Evans, S., Drewry, D.J., Robin, G. de Q. 'Radio echo sounding in Antarctica, 1971–72'. *Polar Record*, 1972, **16** (101) 207–212.
[78] Evans, S., Robin, G. de Q. 'Ice thickness measurement by radio echo sounding, 1971–1972'. *Antarctic J. U.S.*, 1972, **7** (4) 108–110.
[79] Fitzgerald, W.J., Paren, J.G. 'The dielectric properties of Antarctic ice'. *J. Glaciol.*, 1975, **15** (73) 39–48.
[80] Fujino, K. 'Electrical properties of sea ice'. In: H. Oura, ed. *Physics of Snow and Ice: Proc. Int. Conf. on Low Temperature Science*, 1966, Vol. 1, Part 1, Institute of Low Temperature Science, Hokkaido University, Sapporo, 1967, pp. 633–648.
[81] Gassett, R.M. *et al.* 'Bottom topography and gravity, Dome C, Antarctica' (abstract). Presented at AGU Mid-west Meeting, DeKalb, 18–19 September 1980.
[82] Glen, J.W., Paren, J.G. 'The electrical properties of snow and ice'. *J. Glaciol.*, 1975, **15** (73) 15–38.
[83] Goodman, R.H. 'Time-dependent intraglacier structures'. *J. Glaciol.*, 1973, **12** (66) 512–513.
[84] Goodman, R.H. 'Radio echo sounding on temperate glaciers'. *J. Glaciol.*, 1975, **14** (70) 57–69.
[85] Goodman, R.H. *et al.* 'Radio soundings on Trapridge Glacier, Yukon Territory, Canada'. *J. Glaciol.*, 1975, **14** (70) 79–84.
[86] Goodman, R.H., Terroux, A.C.D. 'Use of radio echo sounder techniques in the study of glacial hydrology'. IASH Pub., No. 95 (*Proc. Symp. on the Hydrology of Glaciers*, Cambridge, 7–13 Sept. 1969), 1973, p. 149.
[87] Gow, A.J., Ueda, H.T., Garfield, D.E. 'Antarctic ice sheet: preliminary results of first core hole to bedrock'. *Science*, 1968, **161** (3845) 1011–1013.
[88] Gudmandsen, P. 'Notes on radar sounding of the Greenland ice sheet'. In Gudmandsen, P., ed., *Proc. Int. Meeting on Radioglaciology*, Lyngby, May, 1970. Lungby, Tech. Univ. of Denmark, Lab. for Electromagnetic Theory, 1970, pp. 124–133.
[89] Gudmandsen, P. 'Layer echoes in polar ice sheets'. *J. Glaciol.*, 1975, **15** (73) 95–101.
[90] Gudmandsen, P., Overgaard, S. 'Establishment of time horizons in polar ice sheets by means of radio echo sounding'. Lyngby, Electromagnetics Inst., Tech. Univ. of Denmark, 1978, No. R312.
[91] Hammer, C.U. 'Past volcanism revealed by Greenland ice sheet inpurities'. *Nature*, 1977, **270** (5637) 482–486.
[92] Hammer, C.U. 'Acidity of polar ice cores in relation to absolute dating, past volcanism and radio echoes'. *J. Glaciol.*, 1980, **25** (93) 359–372.
[93] Hargreaves, N.D. 'The polarization of radio signals in the radio echo sounding of ice sheets'. *J. Physics*, Ser. D: Applied Physics, 1977, **10**, 1285–1305.
[94] Hargreaves, N.D. 'Radio echo studies of the dielectric properties of ice sheets'. Ph.D. Thesis, University of Cambridge, 1977.
[95] Hargreaves, N.D. 'The radio-frequency birefringence of polar ice'. *J. Glaciol.*, 1978, **21** (85)

301–313.
[96] Harrison, C.H. 'Reconstruction of subglacial relief from radio echo sounding records'. *Geophysics*, 1970, **35** (6) 1099–1115.
[97] Harrison, C.H. 'Radio echo sounding: focusing effects in wavy strata'. *Geophys. J., Roy. Astron. Soc.*, 1971, **24** 383–400.
[98] Harrison, C.H. 'Radio echo records cannot be used as evidence for convection in the Antarctic ice sheet'. *Nature*, 1971, **173** (3992) 166–167.
[99] Harrison, C.H. 'Radio propagation effects in glaciers'. Ph.D. Thesis, Univ. of Cambridge, 1972.
[100] Harrison, C.H. 'Radio echo sounding of horizontal layers in ice'. *J. Glaciol.*, 1973, **12** (66) 383–397.
[101] Hattersley-Smith, G. 'Results of radio echo sounding in northern Ellesmere Land, 1966'. *Geograph. J.*, 1969, **135** (4) 553–557.
[102] Hattersley-Smith, G., Fuzesy, A., Evans, S. 'Operation Tanquary. Glacier depths in northern Ellesmere Island: airborne radio echo sounding in 1966'. Defense Research Board, Tech. Note 69-6, December 1969.
[103] Heine, A.J. 'Brine in the McMurdo ice shelf, Antarctica'. *New Zealand J. Geol. and Geophys.*, 1968, **11** (4) 829–839.
[104] Hermance, J.F. 'Application of electromagnetic surface waves to the study of the dielectric properties of glacier ice in situ'. In: P. Gudmandsen, ed., *Proc. Int. Meeting on Radioglaciology*, Lyngby, May, 1970, Tech. Univ. of Denmark, Lab. of Electromagnetic Theory, Lyngby, 1970, pp. 84–87.
[105] Hermance, J.F., Strangway, D.W. 'In situ dielectric properties of glacier ice using surface wave interference methods'. Brown University Geophysical Laboratory, Rep. 71-1, 1971.
[106] Jankowski, E.J., Drewry, D.J. 'The structure of West Antarctica from geophysical studies'. *Nature*, 1981, **291** (5810) 17–21.
[107] Jezek, K.C. 'Radar investigations of the Ross ice shelf, Antarctica. Ph.D. Thesis, Geophys. and Polar Res. Center, Univ. of Wisconsin-Madison, 1980, 204 pp.
[108] Jezek, K.C., Bentley, C.R., Clough, J.W. 'Electromagnetic sounding of bottom crevasses on the Ross ice shelf'. *J. Glaciol.*, 1979, **24** (90) 321–330.
[109] Jezek, K.C., Clough, J.W., Bentley, C.R., Shabtaie, S. 'Dielectric permittivity of glacier ice measured in situ by radar wide-angle reflection'. *J. Glaciol.*, 1978, **21** (85) 315–329.
[110] Jiracek, G.R. 'Radio sounding of Antarctic ice'. Geophys. and Polar Res. Center, Univ. of Wisconsin-Madison, Res. Rep. 67-1, 1967.
[111] Jiracek, G.R., Bentley, C.R. 'Velocity of electromagnetic waves in Antarctic ice'. In: A.P. Crary, ed., *Antarctic Snow and Ice Studies II*, American Geophys. Union, Antarctic Res. Ser., 1971, Vol. 16, pp. 199–208.
[112] Johari, G.P., Charette, P.A. 'The permittivity and attenuation in polycrystalline and single-crystal ice 1h at 35 and 60 MHz'. *J. Glaciol.*, 1975, **14** (71) 293–303.
[113] Jones, S.J. 'Radio depth sounding on Meighen and Barnes ice caps, Arctic. Canada'. Canada, Department of the Environment, Inland Waters Directorate, Scientific Series No. 25, 1973.
[114] Keeler, C.M. 'Some physical properties of alpine snow'. U.S. Cold Regions Res. and Eng. Lab., Res. Rep. 271, 1969, 70 pp.
[115] Koerner, R.M. 'Ice thickness measurements and their implications with respect to past and present ice volumes in the Canadian high Arctic ice caps'. *Can. J. Earth Sci.*, 1977, **14**, 2697–2705.
[116] Kovacs, A. 'Radio echo sounding in the Allen Hills, Antarctica, in support of the meteorite field program'. CRREL Spec. Rep. 80-23, 1980, 9 pp.
[117] Kovacs, A., Abele, G. 'Crevasse detection using an impulse radar system'. *Antarctic J. U.S.*, 1974, **9** (4) 177–178.
[118] Kovacs, A., Gow, A.J. 'Brine infiltration in the McMurdo ice shelf, McMurdo Sound, Antarctica'. *J. Geophys. Res.*, 1975, **80** (15) 1957–1961.
[119] Kovacs, A., Gow, A.J. 'Subsurface measurements of the Ross ice shelf, McMurdo Sound, Antarctica. *Antarctic J. U.S.*, 1977, **12** (4) 146–148.
[120] Kovacs, A., Morey, R.M. 'Radar anisotropy of sea ice due to preferred azimuthal orientation of the horizontal c-axis of ice crystals'. *J. Geophys. Res.*, 1978, **83** (12) 6037–6046.
[121] Kovacs, A., Morey, R.M. 'Investigations of sea ice anisotropy, electromagnetic properties,

strength, and under-ice current orientation'. CRREL Rep. 80-20, 1980, 18 pp.
[122] Kuroiwa, D. 'The dielectric property of snow'. *Union Geodes. et Geophys. Intern., Assoc. Intern. d'Hydrologie Sci., Proc. Assemblee Generale de Rome, 1954*, 1956, Vol. 4, pp. 52–63.
[123] Kuroiwa, D. 'Electrical properties of snow'. In: *The Physics and Mechanics of Snow as a Material*, U.S. Cold. Regions Res. and Eng. Lab., Cold Regions Sci. and Eng., 1962, Part 2, Section B, pp. 63–79.
[124] Mae, S. 'Bedrock topography deduced from multiple radar echoes observed on the Mizuho Plateau, East Antarctica. *Antarctic Record*, 1978, No. 61, pp. 23–31.
[125] Maeno, N. 'Measurements of surface and volume conductivities of single ice crystals'. In: E. Whalley et al., eds., *Physics and Chemistry of Ice: Proc. Symp. on the Physics and Chemistry of Ice*, Ottawa, Canada, 14–18 August 1972, Royal Society of Canada, Ottawa, 1973, pp. 140–143.
[126] Maeno, N. 'Investigations of electrical properties of deep ice cores obtained by drilling in Antarctica'. In: *Physical and Chemical Studies on Ices from Glaciers and Ice Sheets*, Monbusho Kagaku Kenpi Sogo Kenkyu (A.), Hokukusho, 1974, pp. 45–56.
[127] Millar, D.H.H. 'Radio echo layering in polar ice sheets and past volcanic activity'. *Nature*, 1981, **292** (5822) 441–443.
[128] Morgan, V.I., Budd, W.F. 'Radio echo sounding of the Lambert Glacier basin'. *J. Glaciol.*, 1975, **15** (73) 103–111.
[129] Neal, C.S. 'The dynamics of the Ross ice shelf revealed by radio echo sounding'. *J. Glaciol.*, 1979, **24** (90) 295–307.
[130] Nye, J.F. 'The distribution of stress and velocity in glaciers and ice sheets'. *Proc. Roy. Soc. London, Ser. A*, 1957, **239**, 113–133.
[131] Nye, J.F. 'Deducing thickness changes of an ice sheet from radio echo and other measurements'. *J. Glaciol.*, 1975, **14** (70) 49–56.
[132] Nye, J.F., Berry, M.V., Walford, M.E.R. 'Measuring the change in thickness of the Antarctic ice sheet'. *Nature Phys. Sci.*, 1972, **240**, 7–9.
[133] Nye, J.F., Kyte, R.G., Threlfall, D.C. 'Proposal for measuring the movement of a large ice sheet by observing radio echoes'. *J. Glaciol.*, 1972, **11** (63) 319–325.
[134] Oswald, G.K.A. 'Investigation of the sub-ice bedrock characteristics by radio echo sounding'. *J. Glaciol.*, 1975, **15** (73) 75–87.
[135] Oswald, G.K.A., Robin, G. de Q. 'Lakes beneath the Antarctic ice sheet'. *Nature*, 1973, **245** (5423) 251–254.
[136] Overgaard, S. 'Dielectric measurements on ice core samples'. Lyngby, Electromagnetics Institute, Tech. Univ. of Denmark., 1981, Report R-86.
[137] Paren, J.G. 'The electrical behavior of polar glaciers'. In: Whalley, E., Jones, S.J., Gold, L.W., eds., *Physics and Chemistry of Ice: Proc. Symp. on the Physics and Chemistry of Ice*, Ottawa, Canada, 14–18 August 1972, Royal Soc. Canada, Ottawa, 1973, pp. 262–267.
[138] Paren, J.G., Robin, G. de Q. 'Internal reflections in polar ice sheets'. *J. Glaciol.*, 1975, **14** (71) 251–259.
[139] Paren, J.G., Glen, J.W. 'Electrical behavior of finely divided ice'. *J. Glaciol.*, 1978, **21** (85) 173–189.
[140] Paterson, W.S.B. 'Temperature measurements in Athabasca Glacier, Alberta, Canada'. *J. Glaciol.*, 1971, **10** (60) 339–349.
[141] Paterson, W.S.B. 'Temperature distribution in the upper layers of the ablation area of Athabasca Glacier, Alberta, Canada'. *J. Glaciol.*, 1972, **11** (61) 31–41.
[142] Paterson, W.S.B., Koerner, R.M. 'Radio echo sounding on four ice caps in Arctic Canada'. *Arctic*, 1974, **27**, 225–233.
[143] Pearce, D.C., Walker, J.W. 'An empirical determination of the relative dielectric constant of the Greenland ice cap'. *J. Geophys. Res.*, 1967, **72** (22) 5743–5747.
[144] Peden, I.C., Rogers, J.C. 'An experiment for determining the VLF permittivity of deep Antarctic ice'. *IEEE Trans. Geoscience Electron.*, 1971, **GE-9** (4) 224–233.
[145] Peden, I.C., Webber, G.E., Chandler, A.S. 'Complex permittivity of the Antarctic ice sheet in the VLF band'. *Radio Science*, 1972, **7** (6) 645–650.
[146] Robertson, J.D. 'Geophysical studies on the Rose ice shelf, Antarctica', Ph.D. Thesis, Geophys. and Polar Res. Center, Univ. of Wisconsin-Madison, Wisconsin, 1975, 214 pp.

[147] Robin, G. de Q. 'Seismic shooting and related investigations'. In: *Norwegian-British-Swedish Antarctic Exp.*, *1949–52*, *Sci. Results 5*, *Glaciology 3*, Norsk Polarinstitutt, Oslo University Press, 1958.
[148] Robin, G. de Q. 'Surface topography of ice sheets'. *Nature*, 1967, **215**, 1029–1032.
[149] Robin, G. de Q. 'Radio echo sounding applied to the investigation of the ice thickness and sub-ice relief of Antarctica'. In: J. Adie, ed., *Antarctic Geology and Geophysics*, Oslo, Universitetsforlaget, 1971, pp. 675–682.
[150] Robin, G. de Q. 'Radio echo sounding: glaciological interpretations and applications'. *J. Glaciol.*, 1975, **15** (73) 49–64.
[151] Robin, G. de Q. 'Velocity of radio waves in ice by means of a borehole interferometric technique'. *J. Glaciol.*, 1975, **15** (73) 151–160.
[152] Robin, G. de Q., Drewry, D.J., Meldrum, D.T. 'International studies of ice sheet and bedrock'. *Phil. Trans. Roy. Soc. London*, 1977, Ser. B., **279**, 185–196.
[153] Robin, G. de Q., Evans, S., Bailey, J.T. 'Interpretation of radio echo sounding in polar ice sheets'. *Phil. Trans. Roy. Soc. London*, 1969, Ser. A, **265** (116) 437–505.
[154] Robin, G. de Q. *et al.* 'Radio echo sounding of the Antarctic ice sheet'. *Antarctic J. U.S.*, 1970, **5** (6) 229–232.
[155] Robin, G. de Q., Swithinbank, C.W.M., Smith, B.M.E. 'Radio echo exploration of the Antarctic ice sheet'. In: Gow, A.J., Keller, C., Langway, C.C., Weeks, W.F., eds., *ISAGE Symposium*, IASH Pub. No. 86, 1970, pp. 97–115.
[156] Rogers, J.C., Peden, I.C. 'VLF electrical properties of the ice sheet measured at Byrd Station'. *Antarctic J. U.S.*, 1973, **8** (5) 241–243.
[157] Rose, K.E. 'Radio echo studies of bedrock in Marie Byrd Land, Antarctica'. In: C. Craddock, ed., *Antarctic Geoscience*, Univ. of Wisconsin Press, 1982, pp. 985–992.
[158] Rose, K.E. 'Characteristics of ice flow in Marie Byrd Land, Antarctica'. *J. Glaciol.*, 1979, **24** (90) 63–75.
[159] Schaefer, T.G. 'Radio echo sounding in western Dronning Maud Land, 1971 – a preview'. *South African J. Antarctic Res.*, 1972, (2) 53–56.
[160] Schaefer, T.G. 'Radio echo sounding in western Dronning Maud Land, 1971'. *South African J. Antarctic Res.*, 1973 (3) 45–52.
[161] Schytt, V. *Glaciology II. The Inner Structure of the Ice Shelf at Maudhein as shown by Core Drilling*, Norwegian-British-Swedish Antarctic Expedition, 1949–52, Sci. Results, 1958, Vol. 4.
[162] Shabtaie, S. *et al.* 'Dome C geophysical survey, 1979–80'. *Antarctic J. U.S.*, 1980, **15** (5) 2–5.
[163] Smith, B.M.E. 'Airborne radio echo sounding of glaciers in the Antarctic Peninsula'. British Antarctic Survey Scientific Report, 1972, No. 72.
[164] Smith, B.M.E., Evans, S. 'Radio echo sounding: absorption and scattering by water inclusion and ice lenses'. *J. Glaciol.*, 1972, **11** (61) 133–146.
[165] Steed, R.H.N., Drewry, D.J. 'Radio echo sounding investigations of Wilkes Land, Antarctica', In: C. Craddock, ed., *Antarctic Geoscience*, Univ. of Wisconsin Press, 1982, pp. 969–975.
[166] Strangway, D.W. *et al.* 'Radio-frequency interferometry – a new technique for studying glaciers'. *J. Glaciol.*, 1974, **13** (67) 123–132.
[167] Swithinbank, C.W.M. 'Radio echo sounding of Antarctic glaciers from light aircraft'. *Proc. IUGG/IASH General Assembly of Berne*, 1967. Commission of Snow and Ice, IASH Pub. No. 79, 1968, pp. 405–414.
[168] Swithinbank, C.W.M. 'Ice movement in the McMurdo Sound area of Antarctica'. *Proc. ISAGE Symposium*, 3–7 September 1968, Hanover, NH, U.S.A., IASH Pub. No. 86, 1970, pp. 472–487.
[169] Swithinbank, C.W.M. 'Glaciological research in the Antarctic Peninsula'. *Phil. Trans. Roy. Soc. London*, Series B., 1977, **279**, 161–183.
[170] Thomas, R.H. 'The distribution of 10 m temperatures on the Ross ice shelf'. *J. Glaciol.*, 1976, **16** (74) 111–117.
[171] Traub, L.T., Gribbon, P.W.F. 'The activation energies of temperate snow samples'. *J. Glaciol.*, 1978, **21** (85) 331–339.
[172] Tupper, W.A., Waddington, E.D., Ricker, K.E. 'Wedgemount Lake and Glacier studies, Northern Garibaldi Park, 1977'. *Canadian Alpine J.*, 1978, **61**, 69–70.
[173] Van Autenboer, T., Decleir, H. 'Airborne radio glaciological investigations during the 1969

Belgian Antarctic Expedition'. *Bull. Soc. Belge de Geologie, de Paleontologie et de Hydrologie*, 1969, **78** (2) 87–100.
[174] Van Autenboer, T., Decleir, H. 'Ice thickness and subglacial relief of the Jelbartisen-Trolltunga area, Dronning Maud Land'. In: Adie R.J., ed., *Antarctic Geology and Geophysics*, Oslo, Universitetsforlaget, 1971, pp. 713–722.
[175] Van Zyl, R.B. 'Radio echo sounding in western Dronning Maud Land, 1972'. *South African J. Antarctic Res.*, 1973, (3) 53–59.
[176] Von Hippel, A.R., Mykolajewycz, R., Runck, A.H., Westphal, W.B. 'Dielectric and mechanical response of ice 1h single crystals and its interpretation'. *J. Chem. Physics*, 1972, **57** (6) 2560–2571.
[177] Waddington, E.D., Jones, D.P. 'A radio echo ice thickness survey of Columbia icefield'. *Canadian Alpine J.*, 1978, **61**, p. 73.
[178] Waite, A.H. 'Ice depth soundings with ultra-high frequency radio waves in the Arctic and Antarctic, and some observed over-ice altimeter errors'. U.S. Army Signal Res. and Dev. Lab., Tech. Rep. 2092, 1959.
[179] Waite, A.H. 'Ice depth sounding by airborne radar'. Paper presented at the Symposium on the Variations of the Regime of Existing Glaciers, Obergurgl, Austria, 1962.
[180] Waite, A.H. 'International experiments in glacier sounding, 1963 and 1964'. *Can. J. Earth Sci.*, 1966, **3** (6) 887–892.
[181] Waite, A.H., Schmidt, S.J. 'Gross errors in height indication from pulsed radar altimeters operating over thick ice or snow'. *Convention Record of the Inst. of Radio Eng.*, March 1961, Part V, pp. 38–54.
[182] Walford, M.E.R. 'Radio echo sounding through an ice shelf'. *Nature*, 1964, **204** (4956) 317–319.
[183] Walford, M.E.R. 'Glacier movement measured with a radio echo technique'. *Nature*, 1972, **239**, 93–95.
[184] Walford, M.E.R., Holdorf, P.C., Oakberg, R.G. 'Phase-sensitive radio echo sounding at the Devon Island ice cap, Canada'. *J. Glaciol.*, 1977, **18** (79) 217–229.
[185] Watt, A.D., Maxwell, E.L. 'Measured electrical properties of snow and glacial ice'. *J. Res. Nat. Bureau of Standards* (Washington, D.C.), 1960, Section D, **64** (4) 357–363.
[186] Watts, R.D., England, A.W. 'Radio echo sounding of temperate glaciers: ice properties and sounder design criteria'. *J. Glaciol.*, 1976, **17** (75) 39–48.
[187] Weber, J.R., Andrieux, P. 'Radar sounding on the Penney ice cap, Baffin Island'. *J. Glaciol.*, 1970, **9** (55) 49–54.
[188] Webber, G.E., Peden, I.C. 'VLF ground-based measurements in Antarctica: their relationship to stratifications in the subsurface terrain'. *Radio Science*, 1970, **5** (4) 655–662.
[189] Weeks, W.F., Gow, A.J. 'Crystal alignments in the fast ice of Arctic Alaska'. *J. Geophys. Res.*, 1980, **85** (C2) 1137–1146.
[190] Weertman, J. 'Can a water-filled crevasse reach the bottom surface of a glacier?' IASH Pub. No. 95 (Proc. Symp. on the Hydrology of Glaciers, Cambridge, 7–13 September 1969), 1973, pp. 139–145.
[191] Whillans, I.M. 'Radio echo layers and the recent stability of the West Antarctic ice sheet'. *Nature*, 1976, **264** (11) 152–155.
[192] Whillans, I.M. 'The equation of continuity and its application to the ice sheet near Byrd Station, Antarctica'. *J. Glaciol.*, 1977, **18** (80) 359–371.
[193] Whillans, I.M. 'Ice flow along the Byrd Station strain network, Antarctica'. *J. Glaciol.*, 1979, **24** (90) 15–28.
[194] Wyeth, R.B. 'The physiography and significance of the transition zone between Graham Land and Palmer Land'. *British Antarctic Survey Bull.*, 1977, No. 46, pp. 39–58.
[195] Yoshino, T., Eto, T. 'Radio echo sounding of Antarctic ice'. In: Murayama, M., ed., *Report of the Japanese Traverse Syowa–South Pole 1968–69, Japan. Antarctic Res. Exp., Sci. Rep. Spec. Issue*, 1971, No. 2, pp. 125–130.
[196] Young, N.W. 'Measured velocities of interior East Antarctica and the state of mass balance within the IAGP area'. *J. Glaciol.*, 1979, **24** (90) 77–87.
[197] Macheret, Yu Ya., Luchininov, V.S. 'Interpretation of results of a surface radar survey of temperate mountain glaciers'. *Materialy Glyatsiologicheskikh Issledovanii. Khronika Obsuzh-*

deniya, 1973, **22** 45–56.

[198] Drewry, D.J. 'Radio echo sounding map of Antarctica (~90°E–180°)' *Polar Record*, 1975, **17** (109) 359–374.

[199] Robin, G. de Q. 'Ice shelves and ice flow'. *Nature*, 1975, **253** (5488) 168–172.

[200] Jezek, K.C., and Roeloffs, E.A. 'Measurements of radar wave speeds in polar glaciers using a down-hole radar target technique'. *Cold Region's Science and Technology*, 1983, **8** 199–208.

[201] Paren, J.G. 'Dielectric properties of ice'. Ph.D. Thesis, University of Cambridge, 1970.

[202] Camp, P.R., Kiszenick, W., Arnold, D. 'Electrical conduction in ice'. In Riehl, N. *et al.*, eds. *Physics of Ice*. New York, 1969, Plenum Press, 256–61.

[203] Clough, J.W. 'Electromagnetic lateral waves observed by earth-sounding radars'. *Geophysics*, 1976, **15** (6a) 1126–1132.

[204] Clough, J.W., Bentley, C.R. 'Electromagnetic sounding at Byrd Station'. *Antarctic J. U.S.*, 1970, **5** (4) 110.

[205] Bogorodsky, V.V., Trepov, G.V., Fedorov, B.A. 'On measuring dielectric properties of glaciers in the field'. In P. Gudmandsen, ed., *Proceedings of The International Meeting on Radioglaciology, Lyngby, May 1970*; Technical University of Denmark, Laboratory of Electromagnetic Theory, 1970, 20–31.

CONCLUSION

On reading this book readers will undoubtedly see the advantages and efficiency of radar methods for the study of the Earth's glaciers. The book brings together 13 different aspects in radioglaciology that demonstrate its extensive and multi-disciplinary usefulness in polar investigations, in geophysics, geology, and geography. Each aspect is a new method of study, and each independently can be developed further.

The book is mainly concerned with Antarctic studies. A brief account is given of theoretical studies. Theory, together with the described field studies, shows the Antarctic ice sheet to be a unique and unknown ice relic. Ice sheets extending over huge land masses obscure their geological and geographical structures. Scientists using radio waves have 'X-rayed' the ice, exposing the subglacial topography of the bedrock. Using this powerful tool, they have discovered subglacial mountains and valleys, high plateaus and depressions, ice-buried debris, and moraine. Sub-glacial lakes were also discovered. Radar sounding has made possible the introscopy of glaciers to find internal layers, to detect crevasses hidden by snow bridges at the surface and near the bottom, and to measure the effective temperature of the ice. Radioglaciology has made it possible also to estimate accurately the water content of glaciers and the velocity of their movement. Knowing the stratigraphy of ice, isochrons have been constructed. This leads the way towards detailed investigations of the history of ice sheets, ice shelves, and glaciation in general. The properties of the electric permittivity tensor for ice were investigated.

Theoretical and field studies of electromagnetic wave propagation in glaciers, and radar techniques for measuring the thickness of ice shelves and outlet glaciers, served as the basis for the successful development of the radar sounding of sea ice and freshwater ice, as well as other strongly absorbing media like permafrost, peat, sand, and the like.

The success of the methods, and the resulting new insights into the problems, proved the value and importance of international cooperation in the study of the rigorous polar environment. It is international cooperation in Antarctic studies that makes it possible to formulate and carry out studies on such problems as the interaction between the Southern Ocean and the solid discharge from the continent, and the growth and decay of sea ice and its impact on the interaction processes on the air-sea-ice system.

The solution of these problems not only is important for basic research, it is of crucial importance for a number of practical tasks.

Summing up, it is appropriate to remember a conversation one of the authors (V.V.B.) had with G. de Q. Robin, then Director of the Scott Polar Research

Institute. Dr. Robin contributed greatly to the development of radar methods and their implementation in glaciology. The authors have repeatedly referred to his works. The conversation took place some 10 years ago in Leningrad, when early radioglaciological results were discussed. Dr. Robin exclaimed: 'How very interesting! You simply cannot help being fascinated by the science you serve with all your heart.'

We can only hope that the fascination of polar regions will never cease to attract young scientists.

SUBJECT INDEX

Absorption
  losses 11, 58, 122f.
  temperature 27, 122f.
Activation energy 35, 39, 87

Bernal and Fowler rules 33

Characteristic frequency 36
Creep of ice 27, 30
Crevasses 164, 179, 211, 226f.

Debye dispersion 84, 88, 90
Debye equations 36
Debye spectrum 36
Defects
  diffusion of 36
  ionic 33
  point 33
  orientational 33
Density
  of firn 16, 46
  of ice 1, 16
  of snow 1, 16f., 87
Dielectric constant (permittivity) 11f., 42, 43, 84f.
Digital processing of data 79f.
Doppler effect 62

Electrical properties
  conductivity 11, 36, 39, 40
  Debye dispersion 84, 88, 90
  permittivity 11f., 42, 43, 84f.
Electromagnetic waves
  focusing 11, 43
  losses in ice 11, 48
  polarization 12, 42, 43, 125f.
  velocity 11, 46, 52, 93f.

Firn
  density 16, 46
  transformation into ice 1, 16

Glaciation
  area 3
  variation 6

Glaciers
  ablation 1, 3, 22
  accumulation 1, 3, 22
  active layer 22
  cirque 2
  crevasses 164, 179, 226f.
  distribution 4
  equilibrium profile 22
  internal melting 223
  introscopy 59
  isotopes in 19
  mass 2, 6
  metamorphic layer 20, 39
  models 13
  mountain 2
  movement 27f., 189f.
  outlet 2
  reflection losses in 41, 55
  retreat of 5
  solid discharge 3
  of Soviet Arctic 5
  strain rate 191, 198
  stratification 20f., 59
  surging 21, 30
  temperature 22f.
  theoretical 29, 234, 236f.
  valley 2
  velocity
    horizontal 28, 189f.
    'instantaneous' 192, 193
    vertical 29, 206f.

Hydrogen bond 32

Ice
  creep 27, 30
  defects 33
  deformation 27, 29, 39
  density 1, 16
  electrical conductivity 11, 36, 39, 40
  electrical permittivity 11f., 42, 43, 84f.
  hydrogen bond 32
  impurities 19, 40

loss tangent 11, 36, 121, 122
polarization 36
strain rate 29
Ice dome 2
Ice sheets 2
  anisotropic layer 113, 114
  crevasses 164, 221, 226f.
  internal melting 223
  internal reflections 107, 117, 164f.
  introscopy 59
  movement 189f.
  reflection losses in 165f.
  temperature gradient, reverse 24
  temperatures 22f., 121
  strain rates 190, 191, 198
Ice shelves 2, 21
Interferometric technique 69, 94

Little Ice Age 6
Looyenga's equation 40, 85
Loss tangent 11, 36, 121, 122

Maxwell's equations 11, 36

Noise temperature 51

Radar sounding of snow 61
Radar systems
  basic equation 48
  noise temperature 51
  sensitivity 56, 57, 58, 59, 60, 61, 62
Reflection losses 11, 41, 48, 55, 165f.
Relaxation time 36, 40, 84

Seismic method 7f.
Snell's law 42
Snow
  density 1, 16f., 87
  metamorphosis 1, 39

Temperature gradient 23, 24, 27
  reverse 24

Velocity
  of electromagnetic waves 11, 46, 52, 93f.
  of glaciers 28, 29, 189f., 206f.